# The Wrong Assumption
# Revolutionary Scientific Theories that
# Shape the Elusive Supernatural World

## Authored by

## Diego Elustondo

# DEDICATION

*To those who have the vision to see beyond perception*

*... Ines, of course, my beloved wife, and my boss, Luiz, for believing in me*

# CONTENTS

# ABOUT THE AUTHOR

Diego Elustondo, Ph.D, was born in Argentina, where he studied engineering and chemistry. In 2001, he moved with his wife to Canada to work as a research scientist, first at the University of British Columbia and then in the private industry. With this book, Dr. Elustondo seeks to provide answers that address both logic and spirituality.

# FOREWORD

In this book, Diego Elustondo explains fundamental scientific concepts in a simple, didactic, and consistent way. More precisely, he explains the evolution of the scientific knowledge (the human understanding of this world, this universe) from its beginnings to the present days. But this is not just an entertaining text of popular science. Elustondo approaches these concepts from such a perspective that shows how science itself leaves room for those events that we can't explain or measure through a scientific method. In other words, he shows that the existence of "*a superior consciousness*", "*supernatural phenomena*", and "*spiritual beliefs*", does not necessarily contradict science. As a skeptic person, this book did not convert me into a believer, but it changed my mind in one thing: now I think that God is not impossible.

Finally, I personally recommend Elustondo to include this brilliant and exquisite fragment as a prelude to his book. It's called *Argumentum Ornithologicum*, and was written by Jorge Luis Borges:

"*I close my eyes and I see a flock of birds. The vision lasts a second or even less; I don't know how many birds I saw. Is the number defined or undefined? This problem involves the question of the existence of God. If God exists, the number is defined, because God knows how many birds I saw. If God doesn't exist, the number is undefined, because nobody was able to count. In such case, I saw less than ten birds (let's say) and more than one, but I didn't see nine, eight, seven, six, five, four, three, or two birds. I saw a number between ten and one, different from nine, eight, seven, six, five, etc. This number, as an integer, is unconceivable; ergo, God exists*".

**Diego Genovese**

# PREFACE

This book offers a scientific approach to philosophical thinking and encourages readers to explore their spiritual beliefs. Using simple stories and analogies, this book demonstrates that one can apply even the most modern and widely accepted scientific theories to interpret the mystical and supernatural.

On the other side of the argument, many contemporary thinkers prefer to believe that spirituality should be kept out of the scientific debate. These thinkers maintain that, in the most extreme cases, believing in the spiritual is either ignorant or delusional; they argue that predictable evidence alone should explain everything that man should know.

Metaphorically speaking, many scientists hold a certain conviction that science and spirituality should be strictly separated by an imaginary wall, which in turn proposes an obvious paradox as many reputable scientists are also spiritual people. Even for the most skeptical scientists in the world, it would not be realistic to assume that each and every person who has spiritual beliefs is necessarily irrational.

This book proposes a much simpler explanation. It proposes that scientists are working under the wrong assumption when trying to validate the source of spiritual beliefs. Perhaps understanding how the universe connects with life requires assumptions that are not fully measurable and predictable for the human observer.

This book argues that, under the right assumptions, enough measurable evidence exists today to provide near-certain proof that science and spirituality are compatible. Any division between science and spiritual elements that does exist has been created by the people who do not have the vision to see beyond predictable facts.

For those who seek a higher purpose in life, this book illuminates the fascinating and revolutionary scientific theories that can provide physical shape to the otherwise elusive concept of the supernatural world. Unlike similar books that explore the connection between science and religion, this book refrains from manipulating science to fit a particular dogma; it simply highlights scientific concepts that may reveal the presence of a higher consciousness.

Perhaps the most distinguishing feature with respect to others is that this book condenses the scientific knowledge currently dispersed throughout many books. The reader will explore classic physics, thermodynamics, electromagnetism, quantum mechanics, the theory of relativity, the theory of the Big Bang, the theory of evolution, molecular genetics, neuroscience, artificial neural networks, statistics, and many other scientific discoveries and theories that have long fascinated the general public.

Unfortunately, the language of science is not always easy to understand. For this reason, this book uses only what technical vocabulary is necessary, without expecting readers to remember or even understand any equations and figures. It is written for readers with all levels of scientific education. For readers interested in learning more about science, bibliographic references are provided.

*Diego Elustondo*

# ACKNOWLEDGEMENT

I would like to thank Bill Deacon and Julia Denton for helping me to start and finish this book. Bill helped me with the initial proposal and Julia edited the English of the final manuscript. A special thanks to Andres Tanasijczuk and Marta Kulis for writing my biography.

I would also like to thank Melissa Suran, a journalist currently pursuing her doctorate at the University of Texas at Austin. In 2010 Melissa wrote an article called *The separation of church and science: Science and religion offer different worldviews, but are they opposite or complementary?*, and now for this book she wrote:

> *"Science and religion may seem like an unlikely pair, but they tend to end up in many conversations together. While some believe both concepts are synonymous, others see them as different as apples and oranges. The constant banter between logic and belief never ceases, leaving many questioning why both can't coexist. Even though strict Bible-thumpers demand that creationism be taught in lieu of the scientific alternative – evolution – some of the top scientists are arguing that science and religion can live in harmony. Scientists, like the U.S. National Institutes of Health director Francis Collins, connect with a higher power through the realization that morality, evolution, the structure of the universe, and life in general are more complex than any mortal being could ever imagine. And while religion and spirituality may not be for everyone, such examples in the world of science and religion make it clear that they can peacefully exist side by side".*

2

# CHAPTER 1

## On the Separation of Church and Science

**Abstract:** "On the Separation of Church and Science" provides an introduction to the book. It simply starts the conversation about the controversy between science and spirituality. This chapter uses the constitutional principle of separation of church and state as a metaphor for how modern secular societies may perceive spirituality as the antithesis of science. It claims that many people today feel comfortable with the idea that science and spirituality should be two separate areas of knowledge; however, the author personally disagrees with this notion. If the right assumptions are made, then science can be used to validate invisible elements of the universe that could provide tangible evidence of the presence of a higher consciousness.

**Keywords:** God, the bible, creationism, theory of evolution, separation of church and state, conflict between science and religion, creation science, intelligent design, religious indoctrination, state atheism, the scientific revolution, the galileo affair, philosophers, self-evident truth, senses of perception, empirical validation, the scientific method, scientific facts, spiritual beliefs, higher consciousness.

## IN THE BEGINNING

At the start of the 20th century, a high school teacher in the United States was arrested for teaching the theory of evolution (The State of Tennessee *vs.* Scopes, 1925). Many people at that time believed that evolution was a reasonable idea, but it was rejected by others because it contradicted the literal interpretation of the Bible; that God created man in his own image (Genesis 1:27).

If one is honest regarding historic facts, then it would be unfair to say that creationism was unreasonable. As long as people could remember, everybody had a father and mother; unless people had existed forever, there had to be a first man and woman who were created by something else. The Bible provided an explanation that was too obvious to be dismissed.

In addition, creationism also provided humans a spiritual purpose of life. If the theory of evolution was correct, then the only purpose of life on earth would have been to survive and reproduce. It is not difficult to understand why people who sincerely believed in the sacred purpose of life may have felt threatened by the theory of evolution.

**Figure 1:** Photograph of John Scopes from the Smithsonian Institution, 1925

Certainly, religious people did not want their children learning in public school that their existence was merely based on the survival of the fittest. That was perceived as a threat not only to religion but also to human decency. For example, a very powerful leader wrote during that same year that "*those who do not want to fight in this world of eternal struggle do not deserve to live*" (Adolf Hitler, 1925).

In response to the anxiety of such an idea, a religious organization successfully promoted state legislation that made the teaching of evolution illegal in public schools.

**THE MONKEY TRIAL**

Of course, not everybody was eager to ban evolution from schools. A civil liberties union decided to challenge this prohibition by asking a high school teacher to intentionally break the law. This legal case probably started as a defense of people's rights and freedoms, but it quickly evolved into a national debate regarding the conflict between science and religion.

At first, the defense tried to demonstrate that there was no actual conflict between science and religion, but gradually the debate turned into a rhetorical battle between the theory of evolution and the literal interpretation of the Bible. Unfortunately, for the argument for science, many questions still remained for which the theory of evolution could not provide conclusive answers.

When the people of the jury had to decide whether the teaching of evolution was right or wrong, they simply declared that it was not within the incumbencies of the court to decide on matters of science and religion. In the end, the teacher was found guilty of breaking the law and ordered to pay a $100 fine.

## THE SEPARATION OF CHURCH AND STATE

Fortunately, the United States Constitution specifically instructed the government to pass no law regarding the establishment of a religion (First Amendment, 1791). One of the first American presidents illustrated this mandate using a metaphor (Thomas Jefferson, 1802); he said that the whole people declared that their legislature should build a wall of separation between church and state.

It was only a matter of time until the laws prohibiting the teaching of evolution were going to be deemed unconstitutional according to the principle of separation between church and state. Effectible, this happened a few decades later after another high school teacher decided to challenge her government (Epperson *vs.* Arkansas, 1968). She challenged a state statute that did not allow the teaching of evolution in public schools and universities.

By that time, however, there was overwhelming support for science, and the jury had no other option but to accept that creationism was an element of religion. The Supreme Court of Justice finally concluded that prohibiting the teaching of evolution was unconstitutional.

## CREATION SCIENCE

The court may have allowed evolution back into the classroom, but it did not end the controversy between science and religion. Some people proposed a new interpretation of the Bible called creation science (John Whitcomb & Henry Morris, 1961). Creation science was still an advocate for religion, but it assumed that the literal interpretation of the Bible could also have scientific explanations.

If this new theory had been true, then creation science should have been taught in public schools as a valid alternative to the theory of evolution. Apparently it wasn't, and creation science was quickly discredited by scientists. It was probably too obvious that creation science was not proposed with the intention of enhancing public education.

In a little more than 20 years, at least eight court rulings were made against state laws or school regulations declaring that the teaching of creation science was unconstitutional as an alternative to evolution (The National Center for Science Education, 2007).

## INTELLIGENT DESIGN

With creation science definitely removed from school, some people during the late 1990s came up with an even newer theory called intelligent design (Percival Davis & Dean Kenyon, 1989). This theory did not explicitly reject evolution or support religion, but tried to demonstrate that the process of evolution had to be partially controlled by a supernatural intelligence.

Because intelligent design did not say who this supernatural intelligence was, its supporters hoped that it would be immune to the separation of church and state. These hopes, however, vanished in the early 2000s when a school was found guilty in court of recommending students to read about intelligent design (Kitzmiller *et al. vs.* Dover, 2005).

In this case, the judge explicitly declared that intelligent design was not science, because it "*violates the centuries-old ground rules of science by invoking and permitting supernatural causation*" (Judge Jones III, 2005). The school was found guilty of crime and ordered to pay more than $1,000,000 in fees.

## THE SEPARATION OF CHURCH AND SCIENCE

Apparently, the long-lasting controversy between science and religion was finally resolved through law. American lawmakers found a way to separate science and religion based on what they thought it should be supernatural causation. If one has to use a metaphor to explain this verdict, then it could be said that the lawyers built a wall of separation between church and science.

Of course, the separating wall between church and science is just a metaphor, but it accurately illustrates how people in modern secular societies understand the role of science and religion. In most modern societies, reasonable people would probably agree that the stories found in religious books are not accurate from a scientific point of view.

Reasonable people would rather believe that science and religion should be two completely independent areas of knowledge. Nobody who defines him or herself as a competent professional would openly invoke supernatural causation to solve practical problems; this would not widely be considered professional at all.

## THE BELIEF IN FAIRY TALES

Science is like the police of secular societies, while religion tries to break the laws of nature. That is, most scientists would never accept any hypothesis that invokes or permits supernatural causation. This in turn proposes an interesting paradox. Many competent professionals around the world do not see any contradiction between their understanding of science and spiritual beliefs.

The United States provides an excellent example of that paradox. It is one of the most scientifically and technologically advanced countries in the world, yet many of its citizens would like to believe that evolution was partially controlled by a supernatural intelligence. According to a recent study, the majority of scientists asked in a survey claimed to be spiritual (Elaine Ecklund, 2010).

It seems evident that, regardless of their level of education, people who are exposed to spiritual beliefs in the early stages of life are very likely to hold spiritual beliefs as adults.

It could be argued, of course, that children in the early stages of life can be indoctrinated to believe whatever the nurturing adult wants. However, this cannot be the only explanation. There is no doubt that children cannot differentiate real stories from fairy tales, but the belief in fairy tales eventually vanishes as children grow up. Children eventually figure out that characters like Santa Claus and the Tooth Fairy are not real, but imaginary.

At some point in life, children realize that their parents lied to them about certain fairy tales in order to make them feel happy. In fact, parents also feel happy to be able to convince their children of the lies they tell them. Nobody who really loves children would honestly believe that the law should protect children from being exposed to fairy tales.

## STATE ATHEISM

With regard to their literal meaning, many spiritual beliefs are no different from fairy tales. If indoctrination were the only reason people held those spiritual beliefs, then it would only be a matter of time before any government would be able to convince its citizens that these beliefs are contrary to the scientific facts.

Indeed, some countries in recent history have already tested that assumption. Countries like France, Russia, and China banned religion from public places with the hope of protecting people from being indoctrinated. Apparently, the experiment has failed. People in those countries are perhaps as spiritual today as they were before their beliefs were banned (Alister McGrath, 2004).

## THE RIGHT TO BELIEVE

It would be too simplistic to assume that people have spiritual beliefs only because somebody told them to have them. It is probably more reasonable to assume that these people were born with the need to find a spiritual purpose in life. Much like an orphan who needs to find the real parents in order to know who he or she really is, some people may have been born with a set of spiritual questions that needed to be answered.

This assumption is also an excellent reason for not allowing religious indoctrination in public schools. If some children are born with the need to find a spiritual purpose in life, then they should have the right to seek for answers in the environment that is better for them and their families. It seems evident that, once these answers are found, most people would find no reason to ask the same questions again.

Under this assumption, the separation of church and state is a human right that protects both believers and unbelievers.

The separation of church and science, on the other hand, makes little sense. It asks believers to choose between two conflicting options that, for some, are both possible. This choice is akin to choosing between being ignorant with a spiritual purpose in life and being educated with a sole purpose of surviving and reproducing.

## THE GALILEO AFFAIR

Most historians would say that the separation of church and science started in the mid-1500s, when an astronomer proposed the theory that the earth was moving in circles around the sun (Nicolaus Copernicus, 1543).

Previously, the general consensus was that the earth was motionless in the center of the universe, while the sun, moon, and planets were moving in circles around the earth. That was the theory proposed by Greek philosophers more than 2,000 years ago (Aristotle's Heaven, ~350 BC), and it was apparently consistent with the literal interpretation of the Bible (Psalm 104:5 & Ecclesiastes 1:5).

When a scientist in the early 1600s showed more experimental evidence that contradicted the ancient theory of the universe (Galileo Galilei, 1610), the church was afraid that this evidence could also contradict the Bible. This resulted in the first formal trial between church and science, and as a prologue for the monkey trial, it ended with the scientists found guilty of crime (Papal Condemnation of Galileo, 1633).

## THE END OF PHILOSOPHY

Just before the Galileo affair, scientists did not really exist; they were rather called philosophers. Philosophers also proposed theories to explain natural phenomena, but they relied on pieces of evidence that were completely imaginary.

Philosophers believed that some self-evident truths could be known by reason alone. For example, the belief in the infinite is often presented as a self-evident truth.

By just looking at the stars at night, the universe for ancient Greek philosophers seemed to be enclosed within a spherical wall. To see the infinite behind that wall, a philosopher had to exercise imagination instead of the senses of perception (Giordano Bruno, 1584). He argued that if the universe is surrounded by a spherical wall, then there has to be something on the other side of the wall. In other words, it was impossible to imagine a sphere big enough that would not have anything on the other side.

The scientific revolution, on the other hand, was defined by the beginning of empirical validation. Empiricists did not believe that knowledge could be achieved through reason alone, but rather from the information provided by the senses of perception. They became convinced that people should not use reason to explain perception, but rather use perception to validate reason.

## EMPIRICAL VALIDATION

A simple example of empirical validation is a man in the desert who sees what seems to be water in the very far distance. It is well known that, when the ground is very hot, an observer may sometimes see an illusion of water in the distance. This is a natural phenomenon known as a mirage – a thin layer of hot air that reflects light in the same way as light reflects off the surface of water.

Since the light reflected by hot air is identical to the light reflected by water, a man using his eyes alone cannot possible determine the object he is seeing. The man can look at the light and interpret the idea of water, but the idea is always imaginary regardless of whether the water is real or not. To validate the idea, the man needs to confirm with his other senses of perception.

If the man starts walking, for example, and the water seems to move farther away as the man walks, then it is clear that the vision is a mirage. If the water does not move, however, it does not necessarily mean that it is actually water. The light could be reflected by a gigantic mirror or by another unexpected liquid, such as mercury or oil.

For further validation, the man needs to take a much closer look at the water. He must more closely see that the liquid is transparent, touch it to verify that it flows, and finally taste it to be sure that it tastes like water. If all of the information provided by the senses is consistent with the idea of water, then the man has no other option but to believe that the water is actually real.

## THE BELIEF IN GRAVITY

The flaw in empirical validation is that it is limited by the senses of perception. The senses of perception are generally classified as sight, hearing, touch, taste, and smell. There are a few more senses that are not included in this list, but as an approximation, this list accounts for the modes of gathering evidence for everything that people can believe through empirical validation.

Clearly, it would be extremely naïve − even arrogant − to assume that the human senses of perception were designed to perceive everything that exists in the universe. It is much more reasonable to assume that the universe exists independent of whether humans can perceive it or not.

In fact, this last assumption is the main difference between empiricism and science. Scientists are empiricists who believe that some elements of the universe are invisible to the senses of perception. Scientists believe in not only what they can see, hear, touch, taste, and smell, but also in other invisible entities that are the products of their own imaginations.

Gravity is probably the most renowned of these entities. Everybody can see that objects spontaneously fall, but the actual entity that attracts them to the ground is invisible to human beings. For centuries scientists and philosophers have used their imaginations to describe and explain gravity. First they imagined it as the manifestation of the need to move towards the center of the universe, then as a force field that propagates through space, and finally as the curvature in the geometry of the universe.

Today, nobody knows what gravity really is, but scientists believe it is real because they can deduce this belief from its effect on our surroundings.

## THE SCIENTIFIC METHOD

To validate concepts like gravity, scientists developed a procedure that is formally known as the scientific method. The scientific method involves using something imaginary to predict the behavior of something measurable. If the measurable behavior is equal to the predictions, then it is said that the imaginary concept is validated with experimental data.

For example, scientists in the late 1600s proposed mathematical equations to predict how objects should move in the presence of other objects (Isaac Newton, 1687). One of these equations was called the universal law of gravity. Then they tested these equations by predicting the trajectory of cannon balls and the orbits of the planets around the sun.

Once they had enough experimental data to validate that the law of gravity was correct, it followed by deduction that there has to be a gravity field connecting the objects and executing the law. Of course, the law of gravity was imaginary, but scientists believed it was real because the objects behaved as if the law of gravity was real.

## THE BELIEF IN SCIENCE

More than 400 years after the scientific revolution, scientists have validated many elements of the universe that are invisible to the senses of perception. Amazingly, some of them are remarkably similar to concepts that were traditionally associated with spiritual beliefs.

Of course, most people do not need any scientific explanation to validate their spiritual beliefs; most people are comfortable with the idea that science and spirituality are separated areas of knowledge. However, this does not necessarily mean that science and spirituality should be separated by an imaginary divider. If there is a supernatural power capable of affecting the lives of spiritual people on earth, then even scientists should be able to feel this power in one way or another.

In order to understand this power, it is good idea to believe in science. Science is valuable to human knowledge, because it expands perception of the natural world into areas that are humanly imperceptible. By using the basic rules of the scientific method, it is possible to perceive many more things in the universe than those that are evident to the naked eye. Even for a total skeptic, there is no good reason not to accept these scientific facts as truth; if the source of spiritual beliefs is real, then it is real for everybody alike.

Many small coincidences can be interpreted and explained in many different ways; thus, any open-minded person should always have a certain degree of philosophical doubt. After all, doubt is what drives scientists to challenge their preconceived beliefs and imagine new elements of the universe that were previously hidden from the senses.

Connecting science and spiritual beliefs, therefore, is like opening a window in the imaginary wall that separates faith from reason, enabling people to explore their mystical thoughts without fear of losing their connection to logic.

## A PERSONAL STORY

Mystical thoughts are not just a product of spiritual beliefs. People may entertain mystical thoughts, regardless of whether they believe them. A premonition, for example, is a mystical vision of sorts that any person may experience. If a person had the absolute conviction that an occurrence was going to take place, and then it does, he or she would find it very difficult to believe that it happened merely by coincidence. For the person that had the premonition, it is much easier to believe that there was an invisible connection between the mind and the circumstances.

To elaborate a little more on this concept, I will refer to a personal story from my teenage years. During a weekend while I was in high school, a few friends and I decided to go camping on the banks of a river. We planned to walk several kilometers upstream along the river, built a wood raft with logs gathered from the area, and then return to our campsite by floating on our raft on the river.

Because our city was located relatively far from that river, we decided to meet at a friend's house located in a town much closer to it. I had never been in that town before, so I asked my friend to wait for me at the bus station so we could go together early in the morning.

Unfortunately, that morning I overslept, and by the time I got to the bus station, my friend had already left. For a short moment, I had the unpleasant feeling that I was going to miss this great adventure that promised to be the time of my life. Because of my lack of planning, I ended up alone at the bus station without any idea of how to find my friends. The only thing I knew was the name of the town.

I quickly became absolutely convinced that I was going to take the next bus and find my friends. For some reason that I still do not understand, I became determined that the bus was going to stop at a place from which I was going to be able to see my friends.

The plan was so clear, that I took the next bus without thinking about it further. The trip took more or less one hour, and when I finally arrived in the town I got off at a random street and start walking in a random direction. I may have walked no more than two blocks when I saw my friends calling to me from the balcony of a house.

According to my friends, they had gone out on the balcony to check if I was coming. Had they not gone out at that exact moment, I would never have known they were in that house waiting for me; I would have continued walking without even noticing that my premonition was fulfilled.

The story finished with another creature waiting for us at the river. On our way back home, floating on a wooden raft, we found a wounded black neck swan that had been shot in the wing. We rescued the swan, and it recovered at home.

Most people would acknowledge the events of this story to be mere coincidence, but from my perspective as an observer, what had actually happened confirmed what I predicted it was going to happen. Even though the odds of finding my friend by coincidence were astronomically low, in my mind I was 100% sure that it was going to happen.

For me there are only two possible explanations for the events of this story; either I was very stupid and extremely lucky, or there was an invisible connection between my mind and my friends. Of course, I do not believe that I had been very stupid.

My question, therefore, is the following: If the second option is possible; would it not be reasonable to explore this connection to enhance my sense of perception? If, on the other hand, scientists prove that the second option is false, can it be possible that scientists are working under the wrong assumption?

## THE WRONG ASSUMPTION

In my opinion, when scientists apply the scientific method to spiritual beliefs, they may be using the wrong assumption.

Scientists assume that, if something is real, then its effect on certain circumstances is measurable and predictable. For example, if I claim that I found my friends by pursuing an invisible connection, then it should be possible to design an experiment in which I find my friends over and over again by using the same connection. My failure to do so would be interpreted as proof that the connection is not real.

However, the source of spiritual beliefs is not necessarily predictable by the human observer. The source of spiritual beliefs is supposed to be a consciousness; thus it could behave differently under the exact same circumstances. Even if the source of spirituality is real, its effect on the circumstances may be decided by an unpredictable consciousness.

Scientists can certainly validate entities that behave as predictable laws, but if an invisible entity exhibits a behavior that scientists cannot predict, then this entity would be also invisible to the scientific method. In other words, spiritual beliefs may be revealed to human beings in unpredictable and mysterious ways.

Perhaps understanding of how the spiritual world connects with human life requires assumptions that are not fully measurable and predictable. If this is hard to believe, then one should reflect on those people throughout history and around the world who were able to convince crowds of civilizations to follow spiritual beliefs that were apparently unreasonable.

Even for the most skeptical scientists in the world, it would not be realistic to assume that each and every person who followed a spiritual leader was either delusional or ignorant. There has to be some kind of logic behind what is seemingly illogical.

## TEARING DOWN THE WALL

It is important to indicate that this is not a religious book. The objective of this book is merely to show that science and spiritual beliefs are compatible. By using non-scientific language and simple analogies, this book intends to create a simplified but clear picture of what spiritual beliefs can be observed through the lens of the scientific method.

After examining the evidence, it would be rather ignorant to claim that spiritual beliefs are not consistent with scientific facts. This is why the principle of separation of church and science is a dangerous fallacy. Inducing people to believe that they have to deny science in order to hold spiritual beliefs can only make them more susceptible to indoctrination.

On the contrary, inducing people to believe that they have to deny spiritual beliefs to understand science can only make them more likely to discriminate against other people just because they think differently. Tearing down the wall of separation between science and spirituality is the best way to protect people from being intellectually manipulated or discriminated against by others.

Elimination of this division can only contribute to the creation of a healthier society in which personal beliefs do not interfere with social relationships. For this reason, this book does not attempt to manipulate science to fit into a particular religion. After describing the facts, this book merely highlights the theoretical entities that seem to validate the mystical and supernatural.

The reader must then reach his or her own rational conclusions. I personally started this book as a mere intellectual exercise, hoping to understand how I could be a scientist while at the same time be inclined to have mystical thoughts. I discovered that the answers regarding my beliefs were always there, within the same science that I thought provided proof against them.

# CHAPTER 2

## The Outer Limits

**Abstract:** "The Outer Limits" seeks to determine if science alone could describe everything that is real in the universe. Most would probably agree that not everything can be explained by science. However, the question addressed here is whether science might be able to explain those things in the future. To answer this question, this chapter quickly moves through history from the discovery of electricity and magnetism to the development of quantum mechanics and the theory of relativity. It concludes that the separation between objects and events only exist within well-defined theoretical limits, and beyond those limits, everything converges into invisible forms of existence. It does not matter how much technology may advance in the future, there are certain things in the universe that can only be perceived in the mind of the observer.

**Keywords:** Units of measurement, electricity, magnetism, theory of electromagnetism, speed of light, special theory of relativity, time dilation, length contraction, general theory of relativity, space time continuum, three-color vision, ultraviolet catastrophe, photoelectric effect, photons, quantum mechanics, the wave function, wave particle duality, uncertainty principle, schrödinger's cat, many-worlds theory.

### THE IMAGINARY WALL

Through perception alone, a person would assume that something is real just because it looks and feels real. The senses of perception, however, only provide a limited amount of information. The senses of perception mainly show light through the eyes, sound through the ears, pressure and temperature through the skin, and a handful of chemical compounds through the nose and the tongue. Everything else is actually perceived by deduction.

When a person views an object, for example, he or she does not see an object at all; he or she sees only the light that was reflected or emitted by the object. The human mind then connects the dots and deduces the idea of shapes and colors. A simple analogy could be the interpretation of the world from the perspective of a blind man.

Of course, it would be very disrespectful to speculate about the visual experience of a blind person, but if it is similar to that of a sighted person with his or her eyes closed, then the world should look more or less like an infinite dark space. Inside this dark space there are not visible shapes and colors; thus the only way to perceive objects is by using one's other senses, such as touching with the hands.

Touching would certainly provide evidence of texture, but this sense does not indicate if these textures are part of the same object. Unless the blind man can scan the entire surface with his hands, he would have to deduce the object from the few places he touched. The blind person probably understands that he imagines most of what he assumes is real, but he would assume that it is real as long as there is no evidence to the contrary.

For example, if the blind man touches something that feels like a brick, then he would probably not assume that the brick was levitating in the air. He would probably assume that the brick was part of a larger wall that was firmly standing on the floor. The blind man does not need to touch each and every possible brick to confirm that the wall is real; he only needs to touch a few bricks and imagine the rest of the wall.

The assumed wall, of course, would contain many more bricks than the ones he touched, but yet it would be much more realistic than the idea of a few bricks levitating in the air. If by some miracle the blind man becomes able to see, then he would probably realize that the wall he sees is different from the wall he imagined.

What the man would probably not realize is that the wall he sees is also imaginary. He had merely assumed it was real because there was not any evidence to the contrary.

## THE CENTER OF THE UNIVERSE

Even though the senses of perception feel absolutely real, the ideas they create are not necessarily realistic. If the theory of evolution is correct, then the senses of perception did not evolve to provide absolute knowledge of the universe; they only evolved to increase the chances of survival and reproduction among the people on earth.

One of the most obvious examples of this idea is the notion that the sun crosses the sky every day. In ancient times, for example, Greek philosophers proposed the theory that the earth was motionless in the center of the universe, while the moon, sun, planets, and stars were moving in circles around the earth (Aristotle's Heaven, ~350 BC). After some modifications were made during the 2nd century AD (Claudius Ptolemy, ~150 AD), the theory became widely accepted, and most people assumed it was real.

As naïve as it may sound, this theory had nothing to do with ignorance or superstition. It is the direct consequence of how people perceive the notions of distance and motion. Distance, for instance, is perceived through binocular vision, which is enabled by two eyes looking at the same object from slightly different positions. This creates the illusion that there are two imaginary lines connecting the eyes with the object, and the angle between these lines indicates the distance to the object.

When distances are too far, however, the eyes became almost parallel, and the angle disappears; thus everything beyond that point looks more or less at the same distance. As perceived through binocular vision, the sky seems to be painted on a semispherical roof that is sitting on the horizon line.

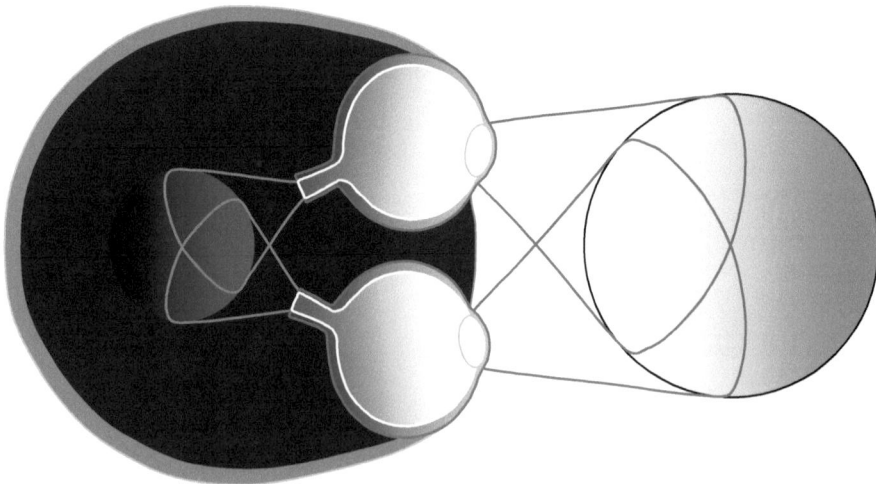

**Figure 1:** My interpretation of binocular vision.

Motion is basically a change in distance; thus it is also perceived through the eyes. Motion, however, is relative to the observer. The eyes cannot distinguish between an object moving toward them or their movement toward the object. In order to determine whether the body is moving, people rely on the ears and skin. Liquid in the ears fluctuates when the body accelerates, and the skin has a sense of touch that can feel the surrounding air flows and determine movement.

Since a person standing motionless on the ground would not feel any acceleration and airflow, then he or she would intuitively assume that the ground is motionless and the sky is moving.

## I THINK, THEREFORE I EXIST

For thousands of years the idea that the earth was at the center of the universe was assumed to be true, but after the invention of the telescope in the early 1600s, scientists found some evidence of the contrary (Galileo Galilei, 1610).

People living at that time and witnessing these discoveries started to doubt everything that they had believed before; if the sun did not cross the sky every day, then any other thing that seems to be real could be only a merely convincing idea. One philosopher even started to believe that only ideas could be proven real (René Descartes, 1637). Assuming that ideas did not exist, he created an idea that contradicted this assumption.

The problem, however, was that ideas were only experienced by the observer. If people had to rely on their own ideas to explain the universe, then there would be as many universes as there are people thinking about it.

## REFERENCE UNITS

To make one person's ideas visible to everybody, the ideas need to be translated into something tangible. For example, if a man is thinking of a dog, then he can explain the idea by pointing his finger to a real dog. In most cases, however, that exact example is not available, thus the ideas have to be broken up into simpler concepts. These simple concepts in science are called the reference units.

A reference unit is basically a standard object or event containing an amount of a fixed physical magnitude. Then, the task of measuring other amounts of the same magnitude consists of counting how many reference units can be accommodated within.

For example, if somebody wants to tell the distance between two marks on the ground, then he or she can use the feet as reference unit. He or she can count how many steps can be walked between the two marks and tell this number as the measure of distance. Everybody else having the same stride can walk the same number of steps and reproduce approximately the same distance.

The immediate advantage of measuring with reference units is that ideas can be translated into numbers. As long as people agree on the reference unit, then everybody thinking about the same number would visualize exactly the same magnitude. In other words, ideas that can be described using scientific units of measurements can exist independent of the first-hand observer.

## THE OUTER LIMITS

Today, scientists have reached the conclusion that everything that is tangible to human beings can be described thoroughly using only seven reference objects and events. They demonstrated that, regardless of how complex the natural world may look to the human eye, it can always be described as the combination of seven basic units of measurement (SI Base Units). The most common units are meter, kilogram, second, and degrees of temperature, but there are also units for measuring electricity, luminosity and number of molecules.

This in turn provides an excellent theoretical background for defining the boundaries of the real world. If people assume that something is real just because it looks and feels real, then they can be sure that everybody sees and feels the same thing by translating it into scientific units of measurements. In the same way, if something can be translated into scientific units of measurements, then it means that this thing looks and feels real to others independently of the observer.

By using scientific units of measurement, scientists can define the boundaries of the real world. Therefore, the question is whether more things exist beyond those boundaries. If there are real elements in the universe that cannot be described with units of measurement, then they exist in forms that humans do not perceive as real.

## THE PHILOSOPHERS' STONES

Light is the cornerstone of human perception, thus it would not be a surprise if it is also the main limitation. The limitations of light can be deduced starting with an examination of two ancient stones called loadstone and amber. The first written records about these stones can be traced back as far as ancient Greece (Thales of Miletus, ~600 BC). Greek philosophers wrote that loadstone, an ordinary rock found in the city of Magnesia, had the natural capability of attracting iron, while amber, which was a translucent gemstone made of fossilized resin, was able to attract feathers and hairs after being rubbed on animal fur.

It took, however, another 2,000 years to understand finally the differences between loadstone and amber (William Gilbert, 1600). By then, scientists realized that loadstone was a natural magnet, while rubbing amber over fur generated an electric charge.

Magnets behaved as if they had two opposite magnetic poles. Each pole attracted the opposite pole of another magnet and rejected the similar one. Iron in close proximity to a permanent magnet became a temporary magnet itself; thus it spontaneously aligned with the permanent magnet and was attracted by the opposite pole.

The first practical application for magnetic poles was probably the Chinese compass invented in the mid-1000s (Wujing Zongyao, 1044). It turned out that the earth was also a gigantic magnet, and a magnetic needle floating over the surface of water spontaneously aligned in the direction of the earth's poles.

With regard to the amber, scientists deduced that, when two non-metallic materials touched, they exchanged a substance called electricity. Then, the excess of electricity in one material attracted the lack of electricity in the other. The attraction ended simply when the electricity returned to the original place, in many cases through the discharge of sparks.

The first practical application of electric sparks was probably the friction machine invented in the mid-1600s (Otto von Guericke, 1663). The friction machine was basically the mechanical equivalent of rubbing amber on animal fur. It generated a constant flow of sparks by pressing a fur-like pad over the surface of a rotating sphere.

Interestingly, the earth also turned out to be a gigantic friction machine that generated lightening by rubbing clouds over its surface. To prove it, a scientist attached a piece of metal to a kite and flew it during a thunderstorm (Benjamin Franklin, 1752). He thought that metal attracted electricity, so he also tied a key to his end of the string. As soon as lightning struck the kite, he felt the sparks jumping from the key to his hand.

## THE ELECTRIC BATTERY

In the late 1700s, scientists discovered by accident that electricity could also make dead bodies to move (Luigi Galvani, 1791). According to the story, a scientist used a knife that had been in contact with an electric charge to cut a dead frog in half. To his surprise, as soon as the knife cut through the frog's spine its legs suddenly kicked back as if the frog were alive.

These findings were then confirmed by inserting a metallic cable into a dead frog and touching the cable with an electric charge. In one of these experiments, however, another accident happened. The scientists touched both the frog and the cable with a different metal and the frog kicked back again without an electric charge. He called this phenomenon animal electricity, because he thought that the frog had generated the electricity inside its body.

A decade later, scientists found a much better explanation for animal electricity. They demonstrated that the electricity was not generated by the frog, but rather by a chemical reaction between two different metals connected through an acidic solution. They demonstrated that it was possible to generate the same electricity by connecting the metals through paper soaked in acid (Alessandro Volta, 1800). This new method to generate electricity was eventually called the electric battery.

## THE THEORY OF ELECTROMAGNETISM

In the early 1800s, while showing an electric battery to his students, a scientist accidentally found an invisible connection between magnets and electricity (Hans Ørsted, 1820). He apparently had a compass sitting beside the battery, and as soon as he turned the current on, the magnetic needle of the compass deflected from the direction of North.

Scientists immediately proposed a theory to explain that phenomenon. They proposed that electricity flowing inside a metal generated a magnetic pole (André-

Marie Ampère, 1820). A decade later scientists also found that the opposite was true. They demonstrated that moving a magnetic pole close to a metal generated an electricity flow (Michael Faraday, 1831).

The results showed that those legendary forces generated by loadstone and amber were actually connected through their rate of change; that is, the changes in the electric force generated a magnetic force, and the changes in the magnetic force generated an electric force.

Scientists then studied the mathematics of this connection and deduced a general theory of electromagnetism (James Maxwell, 1862). In particular, they demonstrated that, when the changes were cyclical, electric and magnetic forces alternatively created one to each other and propagated through space in the form of waves.

## THE AETHER

Much before electromagnetic waves were discovered, scientists had two competing theories to explain the properties of light. Some scientists believed that light was made of little particles emitted in all directions (Isaac Newton, 1675) while others believed that light behaved like water waves propagating over the surface of water (Christiaan Huygens, 1678).

By the early 1800s, the theory of waves became more reasonable, as it explained the phenomenon of interference (Thomas Young, 1803). Interference occurred when two identical light beams were superimposed at a certain angle, thus creating patterns of lights and shadows that resembled the superposition of water waves.

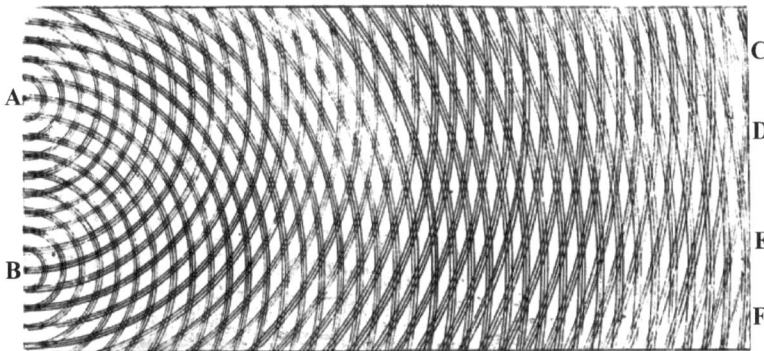

**Figure 2:** Thomas Young's sketch of two-slit diffraction of light, 1803.

The theory of waves, however, presented a major theoretical problem. Unlike water waves, which were obviously made of water, light waves were able to propagate through space. There had to be some kind of invisible substance filling the space that could support the propagation of waves. Scientists called this substance the aether, which was the fifth element in ancient Greek philosophy that filled the universe above the sky.

## THE SPEED OF LIGHT

After electromagnetic waves were discovered in the mid-1800s, they became the perfect candidate as the substance of light. Scientists proved this hypothesis by using speed as evidence. They demonstrated that the speed of light measured experimentally (Hippolyte Fizeau, 1849) was almost identical to the speed of electromagnetic waves calculated theoretically (James Maxwell, 1862).

At first, however, electromagnetic waves still needed the aether to propagate through space. Scientists believed that electric and magnetic forces were physically connected through electricity, thus they still needed some kind of invisible substance filling the space that could support localized electricity flows.

This in turn created a theoretical inconsistency, since electromagnetism predicted that the speed of light was equal in all directions. On the contrary, if the waves were supposed to move through the aether, then light should look faster or slower depending on whether the observer was moving toward or away from the source.

A simple analogy involves two persons on the beach looking at the ocean waves. If one person suddenly starts running towards the ocean, then he or she should see the waves coming faster than the other person, who is standing motionless on the beach.

Since the earth is spinning and moving in orbit around the sun, scientists expected to find huge differences in the speed of light depending on the direction it was measured. Surprisingly, the speed of light was constant. Scientists always failed when they tried to measure different speeds of light in perpendicular directions (Albert Michelson & Edward Morley, 1887).

## TIME DILATION AND LENGTH CONTRACTION

Both theory and practice suggested that the speed of light was constant. However, this was against common sense. In the analogy of the ocean waves, for example, this would have meant that a person running toward the ocean and a person standing motionless on the beach would have both seen the waves coming exactly at the same speed.

If life were a motion movie, then one possibility could have been that the scene of the person running towards the ocean was projected in slow motion. The other possibility could have been that the scene of the person running towards the ocean was projected on a smaller screen. In this case, the person running would have touched the waves first, but because the distance was shortened, the apparent speed could have been the same.

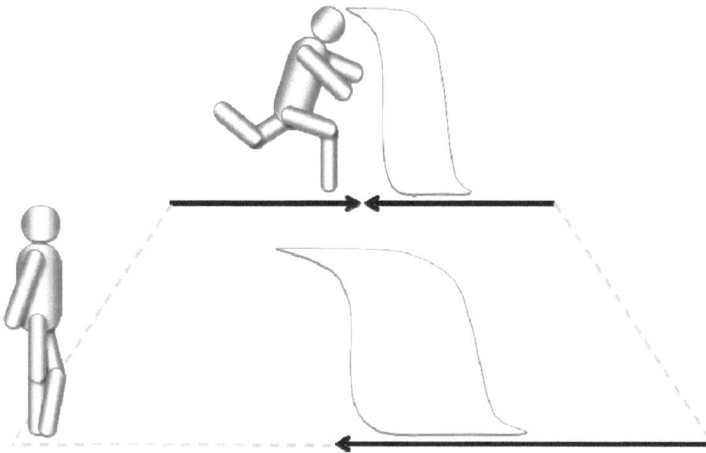

**Figure 3:** My interpretation of length contraction in the analogy of ocean waves.

Surprisingly, this is more or less what scientists did − they proposed a set of mathematical transformations that made this movie-like illusion occur in real life. Basically, they assumed that if somebody moved with respect to the aether, then time slowed down and distances shrank in the precise amounts that made the speed of light to appear constant (Hendrik Lorentz, 1899).

This, of course, was a theoretical transformation. Even though some scientists believed that length contraction could be real phenomenon, most of them agreed that time dilation was probably a mathematical illusion. In addition, the transformation still required the existence of a motionless aether in which time and distances were real.

## THE SPECIAL THEORY OF RELATIVITY

In the early 1900s, one scientist proposed a much more logical assumption to explain time dilation and length contraction. Instead of artificially manipulating length and time to make the speed of light appear constant, he assumed that the speed of light was constant, because it was a law of nature. Then, because the speed of light was constant, time dilation and length contraction were the natural consequences of using electric and magnetic forces to connect matter in motion. This interpretation is known today as the special theory of relativity (Albert Einstein, Sep 1905).

The main limitation of the special theory of relativity was the law of gravity. In classic physics the gravity field was supposed to reach instantaneously all places in the universe, while force fields in the special theory of relativity were not supposed to move faster than the speed of light. To solve this problem, scientists had to come up with a totally new interpretation of gravity.

## THE GENERAL THEORY OF RELATIVITY

In the early 1900s, scientists reinvented gravity for the third time in history. They basically reinterpreted the reason why objects accelerated as they fell down by gravity. Scientists figured out that it was impossible to differentiate whether objects moved faster because they were attracted by a force or if they seemed to move faster because distances were getting shorter and times slower. In other words, gravity may not have been a force, but a curvature in space and time.

This interpretation is known today as the general theory of relativity (Albert Einstein, 1916). A typical analogy can be found by imagining a heavy ball placed over an elastic membrane. If one imagines that the heavy ball creates a deep depression on the elastic membrane, then it is easy to visualize how other smaller balls would fall down into that depression as if they were attracted by the heavy one.

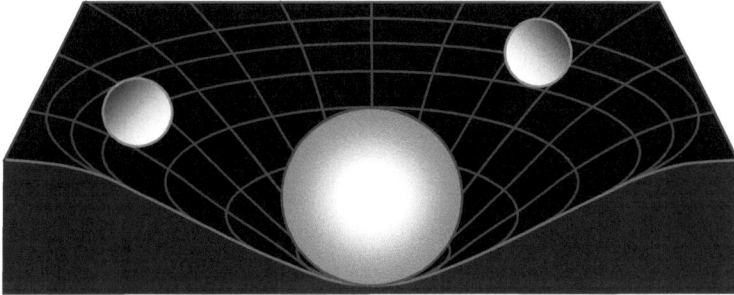

**Figure 4:** My version of the well-known analogy to illustrate space-time curvature.

One famous validation of this theory was based on photographs of a solar eclipse (Sir Arthur Eddington, 1919). The photographs showed that, when the moon momentarily covered the sun, the stars seemed to be farther from the sun than they were supposed to be. This suggested that the light bended due to a space-time depression around the sun.

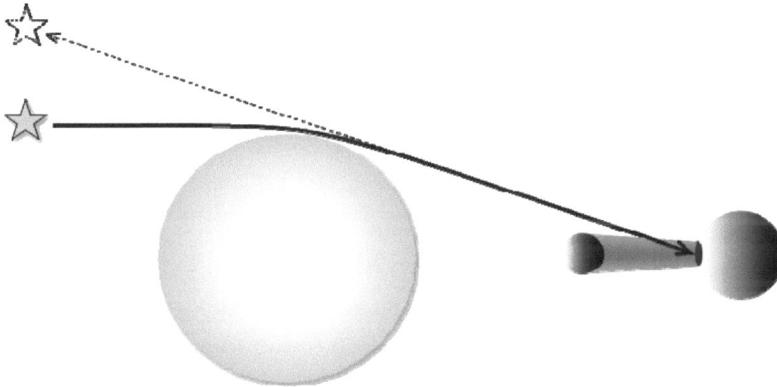

**Figure 5:** My interpretation of light bending around the sun during an eclipse.

Many years later, scientists also validated the concept of time dilation. They compared clocks placed in commercial airplanes with an identical clock placed on the ground (Joseph Hafele & Richard Keating, 1971). Of course, the clocks showed time differences consistent with the theory of relativity.

## THE SPACE-TIME CONTINUUM

If the theory of relativity is correct, then the perception of space and time is not independent of the observer. People can measure space and time and believe that they are real, but other people measuring the same things may get different results. It does not matter how much technology may advance in the future; electric and magnetic forces can simply not move faster than the speed of light and connect everything simultaneously.

If it were possible to conclude this half of the chapter with a metaphor, then one could say that it is theoretically impossible to imagine a wall surrounding the real world. If everything that is defined as real can be described with scientific units of measurement independently of the observer, then units of measurement should be also independent of the observer. Amazingly, scientists demonstrated that without an observer, units of measurement such as distance and time do not even exist as independent entities.

## THE COLORS OF THE RAINBOW

Beside the speed of propagation, the other two basic properties of waves are amplitude and wavelength. If waves are imagined as a series of humps moving one after another in a straight line, then the height of the humps is called amplitude and the width of the humps is called wavelength.

In the case of light, the amplitude is perceived as brightness, and the wavelength is perceived as color. The perception of color, however, is not a direct measure of wavelength. In sound waves, for example, a gradual increase in wavelength from narrow to wide is perceived as a gradual pitch reduction from high to low. In light waves, on the other hand, a gradual wavelength increase from narrow to wide is perceived as a shift through the colors of the rainbow.

The actual relationship between wavelengths and colors was explained in the mid-1800s by the theory of three-color vision (Thomas Young & Hermann von Helmholtz, 1850). According to this theory, the human eye divides light waves in three groups: a group with wide wavelengths that it perceives as red, a group with medium wavelengths that it perceives as green, and a third group with narrow wavelengths that it perceives as blue.

These groups, however, considerably overlap, and some waves are found inside two or three groups at the same time. Consequently, some colors are perceived as a combination of blue, green, and red. For example, a wave that is found inside the red and green groups is perceived as yellow, and a wave that is found inside all the three groups is perceived as white. Because of three-color vision, modern TV screens can create almost any color by using only red, green, and blue dots.

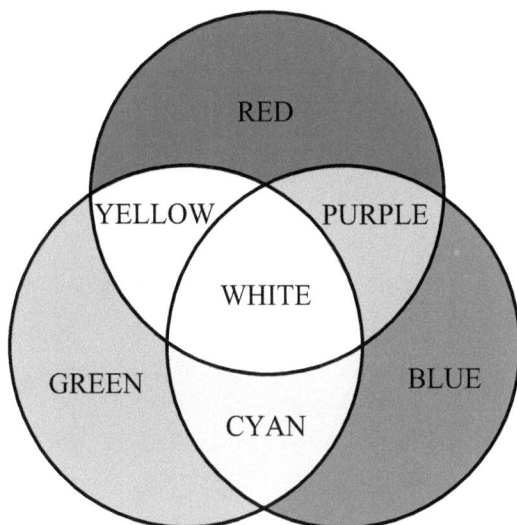

**Figure 6:** My interpretation of the theory of three-color vision.

## THE LIGHT BULB

White light is the best light for seeing the world because it contains all the colors of the rainbow. White light is naturally emitted by the sun and also from materials that are at extremely high temperatures.

After the invention of the electric battery in the early 1800s, scientists quickly discovered that it was possible to create white light by heating a filament with electricity. They also learned very quickly that these filaments burned at high temperatures; thus, scientists placed them inside glass bulbs to protect them from the air. After almost 100 years of trial and error, scientists finally patented the first commercially viable light bulb in the late 1800s (Thomas Edison, 1879).

## THE ULTRAVIOLET CATASTROPHE

One practical problem for light bulb manufacturers in the beginning was the task of generating the brightest white light possible with the lowest electricity consumption. They knew that the light was white when the temperature was very high, but it gradually changed to red when the temperature cooled down. It was clear that there was a direct relationship between temperature and color.

To explain this relationship, some scientists applied the theory of electromagnetism (Lord Rayleigh & Sir James Jeans, 1905). They imagined the molecules inside the materials as little electric oscillators that emitted electromagnetic waves.

The problem was that these oscillators were supposed to be constrained to a certain space. They only had a number of possible ways to generate waves, and this number decreased if the waves were widened. If light was really made of electromagnetic waves, materials should not emit wide waves in the gamma of red but rather narrow waves in the gamma of ultraviolet when the temperature cools down.

Another scientist came up with a completely different interpretation. He assumed that light was emitted in particles (Max Planck, 1900). His idea assumed that the amount of light that could be accommodated inside a particle was reduced if the waves were wider. Consequently, red light was easier to emit because its particles were smaller.

## THE PHOTONS

The theory of light particles worked well to explain the relationship between temperature and color, but nobody believed that these particles were real. Even the idea's creator believed that it was a convenient mathematical illusion. A few years earlier, however, scientists had also discovered the photoelectric effect (Heinrich Hertz, 1886). The photoelectric effect was basically a light bulb working in reverse; it generated electricity by exposing metal to light.

The mystery was that the photoelectric effect did not work for all colors. Scientists found that it was relatively easy to generate electricity with ultraviolet light, but it was almost impossible to do it with red light, regardless of how bright it was.

The answer again laid in the theory of the light particles (Albert Einstein, Jun 1905). Since the amount of light in a particle decreased if the waves became wider, then red particles were simply too small to push electricity out of a metal.

Once the theory of light particles was accepted, scientists started calling them photons.

## QUANTUM MECHANICS

The theory of photons created an interesting paradox related to the properties of light. As far as scientists could tell, light behaved as particles when it was emitted and absorbed by matter, but it behaved as waves when it propagated through space. The answer, they would eventually learn, was hidden inside the properties of matter.

The first clue came from a scientist who shot a thin layer of gold with radioactive particles (Ernest Rutherford, 1911). He found that most particles passed throughout gold as if there was nothing in front of it, but a few particles were deflected by the gold as if they had hit something massive. To his understanding, the atoms of gold seemed to have a massive nucleus in the core and then a cloud of electric particles revolving in orbits around it.

Since pure atoms only emitted or absorbed pure colors of light, scientists became convinced that their orbits should be also fixed (Niels Bohr, 1913). They assumed that the electric particles emitted or absorbed photons when they jumped from one fixed orbit into another. Then, to explain why the orbits were fixed, scientists assumed that any particle in motion could be also imagined as the propagation of a wave (Louis de Broglie, 1924).

The idea was that the electric particles only had a number of possible ways to accommodate their waves within the atoms, meaning that they could only move through fixed wavy patterns. Almost immediately, scientists proposed a mathematical equation to describe those wavy patterns (Erwin Schrödinger, 1926). This mathematical equation, which is known today as the wave function, indicated the probability of finding particles moving within the atoms.

## THE UNCERTAINTY PRINCIPLE

The wave function explained why light behaved as waves and particles at the same time, but it proposed a much more complicated philosophical problem:

matter also behaved as waves and particles at the same time. The solution to this paradox was provided by a scientist who, paradoxically, did not believe in the wave function (Werner Heisenberg, 1927).

He created a complicated matrix model to connect properties that could be known about an atom, and he found that it was impossible to know them all simultaneously. He found that there were pairs of properties that could not be known simultaneously with absolute certainly. That is, knowing more about one caused one to know less about the other.

A simple analogy could be the photograph of a moving object. If a person trying to take a photograph of a moving object wanted to take a well-defined picture of the object, as if it were frozen in time, then he or she would have to use a short exposure time. If, on the contrary, he or she used a long exposure time, then the picture would show a blurry image of the object distributed throughout an entire trajectory. It would not matter how advanced the camera was; it is impossible to take a picture of an object and its trajectory simultaneously.

This conclusion is known today as the uncertainty principle. It basically states that the real nature of matter is unknown, but it collapses into waves and particles when it interacts with the observer (Copenhagen interpretation, 1955). It does not matter how much technology may advance in the future; it is theoretically impossible to describe matter with absolute certainty.

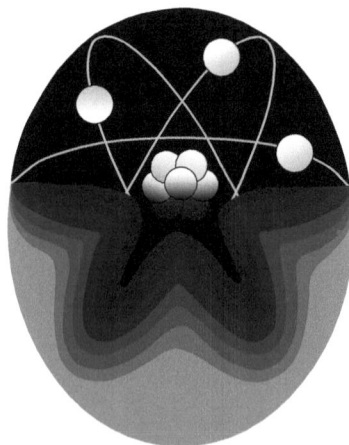

**Figure 7:** My impression of wave-particle duality of matter.

## THE TRUTH THAT LIES BENEATH

If the modern scientific concepts of the space-time continuum and wave-particle duality are correct, then it is theoretically impossible to describe the universe by using only units of measurement. It reaches a point in which the ideas representing the facts can only be seen in the mind of the observer. One can measure something and believe it is real, but someone else can measure the same thing and believe something else.

Thus, it is a myth to believe that measurable facts can be completely separated from ideas.

## SCHRÖDINGER'S CAT

It should be evident at this point that this chapter refrains from speculating about the unknown. It only shows that there are elements of the universe that exist beyond perception.

Scientists, on the other hand, like to speculate about the unknown. For example, some scientists believe that space and time are connected. They found that it is theoretically possible to create a closed loop in the space-time continuum and then literally to travel into the past (Kurt Gödel, 1949).

The most prolific source of speculation, however, is the principle of uncertainty. The typical interpretation of the principle of uncertainty is that reality remains undefined until somebody measures it. Without an observer, all possible forms of reality exist simultaneously as probabilities.

One criticism of that interpretation is a mental experiment known as Schrödinger's cat (Erwin Schrödinger, 1935). The experiment consists of imagining a cat inside a box with a radioactive material that had a 50% chance of emitting a particle and killing the cat. Since nobody is able to see inside the box, then the cat should exist simultaneously as dead and alive.

One possible solution for this paradox is the many-worlds theory (Hugh Everett, 1957). This theory assumes that the different possibilities described by the wave function could be happening simultaneously in parallel worlds. Regardless of whether there is an observer or not, there could be a world in which the cat is dead and another world in which the cat is alive.

Ironically, life after death is a spiritual belief, but a dead cat living in a different world is scientifically possible.

# CHAPTER 3

## The Matter of Facts

**Abstract:** "The Matter of Facts" focuses on the substance that makes up the universe. It explains that, according to the laws of physics, the natural world is entirely made of a substance that can be referred as mass and energy. In the same way, if the supernatural is real, then it also has to be made of something real. The story first looks for this substance in the ancient theory of the four elements and then moves quickly through history, explaining how the concepts of mass and energy were first created in the theory of classic physics and then unified in the special theory of relativity. The theory shows that the universe is either eternal or it was created by something else, and the best explanation of its origin is the theory of the Big Bang. If this theory is correct, then the almost infinite variety of elements that people perceive, including human life and intelligence, comprise only 5% of the mass and energy available in the universe. The rest is invisible to the senses of perception.

**Keywords:** The four elements, square of opposition, center of the universe, universal law of gravity, classic physics, mass definition, principle of mass conservation, mechanic work, energy definition, principle of energy conservation, principle of action and reaction, friction, heat and work equivalence, mass energy equivalence, nuclear fission, atomic bomb, universe expansion, theory of the big bang, dark matter, dark energy.

## THE MATTER OF FACTS

Nobody can deny that the universe contains a vast variety of objects and events. Even if only those things that are within the human reach are taken into account, then there would still be more objects and events in the universe than anything that can be possibly counted during a human life. Common sense would dictate, however, that too many is an unreasonable number. It is easier to believe in a more reasonable number of basic substances, making everything else that is supposed to be real.

Common sense would also dictate that things that are supposed to be real are probably made of something real. It would be very difficult to believe in objects that are made of nothing and events that are driven by no cause. Regardless of whether people can perceive them or not, there is no doubt that real things are made of real substances.

The same principle also applies to spiritual beliefs. If people believe that the source of spiritual beliefs is real, then the source has to be made of something real. In addition, if people believe that the source of spiritual beliefs can affect one's circumstances in life, then the real world and the source must be connected through something real.

Even for those who believe that reality and spirituality should be separated by an imaginary wall, the wall cannot be absolutely impenetrable; otherwise, nobody would ever connect with his or her source of spiritual beliefs.

## THE FOUR ELEMENTS

The most harmonious universe that has been ever been imagined was probably the Greek universe in the $4^{th}$ century BC (Aristotle's Heaven, ~350 BC). For ancient philosophers, the universe was enclosed within a perfect sphere and surrounded by a wall of stars. On the other side of the wall was the prime mover. The prime mover kept the stars rotating around the earth, and this transferred circular motion down into the universe to move the planets, moon, and sun.

In terms of substance, the Greek universe was made of five basic elements. The natural world was made of four elements called earth, water, air, and fire, while the sky was made of a fifth element called aether. In retrospect, earth, water, air, and fire seem to serve as ancient metaphors for the modern concepts of solid, liquid, gas, and energy in the form of heat.

In addition, the ancient elements had also powers. They had three pairs of opposite powers that could make them hot or cold, dry or wet, and ascending or descending. Again, these powers seem to be ancient versions of the modern concepts of temperature, plasticity – because moisture tends to soften things up, and weight.

The elements and their powers were connected through an elegant diagram that is known today as the square of opposition. In the square of opposition, the elements defined the corners of a square, and the powers divided it in opposite halves. Then, by defining clockwise rules on how to read the diagram, Greek philosophers were able to explain the properties and behavior of materials based on the combination of earth, water, air, and fire.

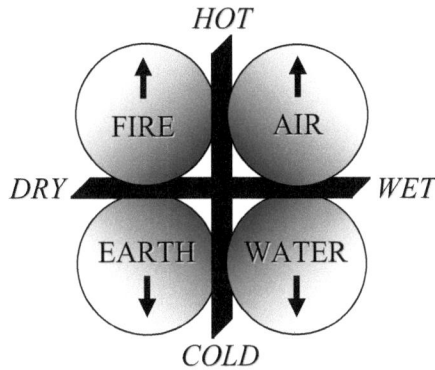

**Figure 1:** My interpretation of the theory of the four elements.

## HORROR VACUI

One practical application of the theory of the four elements was the explanation of gravity and buoyancy. The theory said that the order in which earth, water, air, and fire appeared in the square of oppositions represented the preferential order of these elements with respect to the center of the universe. According to this rule, gravity was the tendency of earth-like objects to move toward the center of the universe, and buoyancy was the tendency of air-like things to be placed above the water.

Ancient philosophers also realized that applying a force would make an object to move, and it was harder to move this object through water than it was to move the same object through the air. With this in mind, they proposed the first law of motion that is known to humankind. They proposed that the speed was equal to the force divided by the weight of the medium (Aristotle's Physics, ~350 BC).

The law of motion was then used to explain another empirical principle that is known today as nature's horror of vacuum. Philosophers realized that vacuum had zero weight, thus a force divided by the weight of a vacuum would have been mathematically equal to an infinite speed. Since infinite speed did not seem to be a reasonable number, they reached to the conclusion that vacuum could not exist.

A simple example of nature's abhorrence of a vacuum could be found with a person drinking water through a straw. When the person sucks water from the top of the straw, he or she creates the vacuum, which nature abhors; thus more water needs to rise from the bottom at infinite speed to fill the vacuum before it has the chance to exist.

## EUREKA

Ancient Greek philosophers also deduced an equation to connect buoyancy and force. The story relates that a king became suspicious about his crown because he thought it may not have been made of solid gold. If the gold had been mixed with another metal, then it would have been impossible for him to prove it without damaging the crown first.

The solution came to the mind of a philosopher while he was taking a bath (Archimedes of Syracuse, ~250 BC). He noticed that when he was submerged in water, the water level rose and his body felt weightless. This convinced him that water was competing with his body to move toward the center of the universe, thus creating an uplift force equal to the weight of displaced water.

If this story was true, then he could prove whether the crown was made of pure gold by weighing it in water. He only needed to place the crown and its weight in pure gold on the opposite sides of the balance, and then determine if they still had the same weight when they were placed under water.

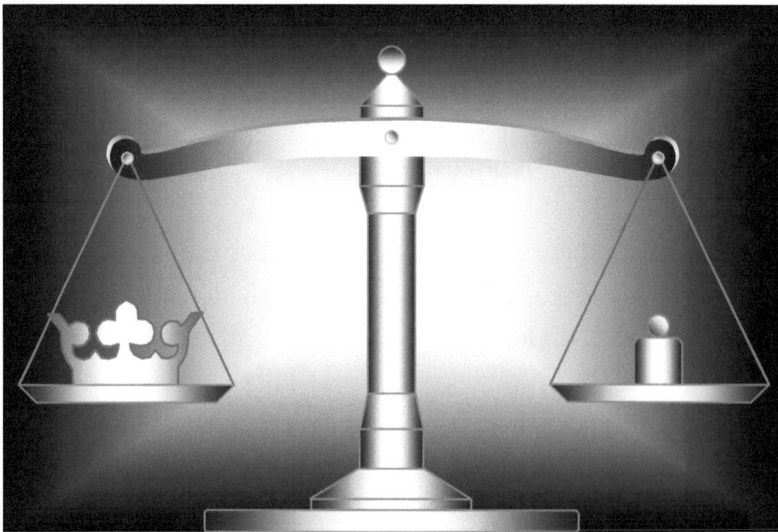

**Figure 2:** My impression of Archimedes' balance.

## THE NEW WORLD ORDER

It is customary to say that the scientific revolution began during the mid-1500s after an astronomer determined that the earth was moving in circles around the sun (Nicolaus Copernicus, 1543). By removing the earth from its former position at the center of the universe, this astronomer removed the fundamental stone upon which the Greek laws of nature were based. It was a matter of time before the ancient notion of the universe would eventually collapse.

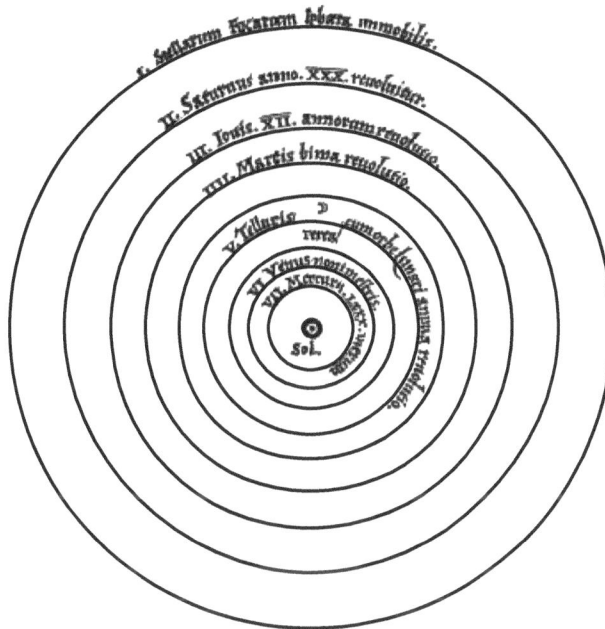

**Figure 3:** Copernicus' eliocentric model of the universe, 1543.

## THE STORY OF THE TWO BALLS

Even though the Greek laws of nature had not been seriously disputed for almost 2,000 years, from the beginning they had an Achilles' heel – they could not explain why, for example, an arrow kept moving after it was shot by a bow. The ancient law of motion stated that the speed was caused by force, but there was no obvious object pushing the arrows from behind.

The solution was to blame nature's horror of vacuum; philosophers assumed that the air pushed the arrow as it rushed to fill the vacuum that the arrow left behind.

In the mid-1600s, the equivalent of the Greek arrow was the cannon ball. Shooting balls with cannons exercised essentially the same concept as shooting arrows with bows, but aiming at the target was much more difficult. The shot had to be aimed to a point in the sky so that the ball hit the target as it came down.

To solve this problem, a scientist came up with a revolutionary empirical method (Galileo Galilei, ~1638). He mentally divided the speed of the cannon ball into two independent directions. He imagined that the ball has one speed parallel to the ground and another speed into the sky.

He first noticed that motion parallel to the ground was the equivalent of rolling a ball over a flat surface; the ball kept moving forward unless there was something to stop it. Contrary to what the ancient law of motion said, the ball did not need a force to move parallel to the ground. Without any force, the ball continued moving at the same speed in which it was moving before. This conclusion is known today as the law of inertia.

In the direction of the sky, on the other hand, motion was the equivalent of throwing a ball up in the air. That is, the ball gradually became slower as it went up, stopped at the peak of the trajectory, and then gradually became faster as it came down. Surprisingly, all balls of different weights became slower and faster at the same rate. This was again contrary to the ancient law of motion that stated that heavier objects should fall faster than lighter ones.

The scientist then tried to prove this finding with a mental experiment. He simply pointed out that regardless of their weight, a heavy cannon ball and a light musket bullet would fall together if they were tied with a string. The popular story, however, tells that this scientist proved his theory by throwing the cannon ball and the musket bullet from the Leaning Tower of Pisa (Galileo Galilei, 1590). If the story was correct, then the two balls hit the ground exactly at the same time.

## THE END OF HORROR

Coincidentally, as soon as the ancient law of motion was proven wrong, nature's horror of vacuum mysteriously started to fail. In the early 1600s, water pumps performed much as a person sucking water through a straw − they were basically metal pipes with a piston inserted in the top. In theory, pulling up the piston created the vacuum that sucked water from below, while pushing the piston down opened a valve that let this water flow.

As technology advanced, these pumps were able to raise water higher, but there was a mysterious height of approximately 10 meters at which all water pumps would stop. At that particular height, nature no longer abhorred vacuum.

To explain this mystery, a scientist in the mid-1600s came up with a vision. He imagined that the atmosphere was the equivalent of an ocean of air (Evangelista Torricelli, 1643). If one could imagine him or herself under a real ocean of water, then it would be easy to imagine that this ocean has weight, and to infer that the weight of the ocean pushes the water into the empty voids.

In the same way, when scientists imagined the world under an ocean of air, they realized that nature's horror of vacuum was just the weight of the atmosphere pushing air into the empty voids. Water pumps just stopped working because the weight of the atmosphere was equivalent to the weight of 10 meters of water.

## THE NEW LAW OF MOTION

By the late 1600s, the findings that started the scientific revolution were mathematically translated into the laws of classic physics (Sir Isaac Newton, 1687).

The development of classic physics is usually associated with the law of gravity. A popular story relates that a scientist came up with the universal law of gravity after being hit in the head by an apple that fell from a tree (Isaac Newton, 1660s). He apparently deduced that, if gravity was able to reach the top of a tree and attract an apple toward the earth, then it might also be able to reach deep into the universe and attract the planets toward the sun.

The most important law of classic physics, however, was the new version of the law of motion, which stated that forces did not cause objects to move, but rather changed the speed at which objects were moving. The ratio between force and change was eventually called the mass.

## THE DEFINITION OF MASS

According to the new law of motion, mass was just a number indicating how difficult it was to change the speed of an object. Weight, on the other hand, was always perceived as something real. People intuitively assumed that weight was a measure of the amount of material contained in an object. For example, if two coins have exactly the same weight, then people assume that they contain exactly the same amount of metal.

Fortunately the weight was the force of gravity, and the change in speed caused by the force of gravity was proven equal for everybody; for the same location on earth, mass and weight were two different measures of the same number. Scientists only needed to define a standard object, such as a liter of water, and then use its weight as the reference unit for mass.

## THE PRINCIPLE OF MASS CONSERVATION

Before scientists imagined the world under an ocean of air, most people were convinced that gas did not have weight. Like the legendary philosopher in the bath, gas felt weightless in the atmosphere because it was pushed up by the weight of displaced air. This confusion created the illusion that mass could be created or destroyed.

A typical example of destruction of mass is setting a piece of wood on fire. If one were to weigh the wood and the ashes before and after combustion, then he or she would measure that the weight of the wood was much higher than its weight after it had burned. Without a better assumption, he or she would conclude that most of the mass had been consumed.

This controversy ended in the mid-1700s, when scientists performed chemical reactions inside sealed containers (Antoine Lavoisier, 1750s). This enabled them to weigh everything, including the containers, before and after the chemical reactions, and the results indicated that mass was never created or destroyed, but only transformed from one form into another.

## THE DEFINITION OF ENERGY

In the early 1800s, scientists started to simplify the mathematics of the law of motion. They applied a mathematical operation that is known as the integral, and they came up with a new theoretical entity called mechanic work (Gustave Coriolis, 1829). Simply put, mechanic work can be imagined as the mathematical metaphor of a horse pulling a load along the street. It was defined as a force applied through a certain distance.

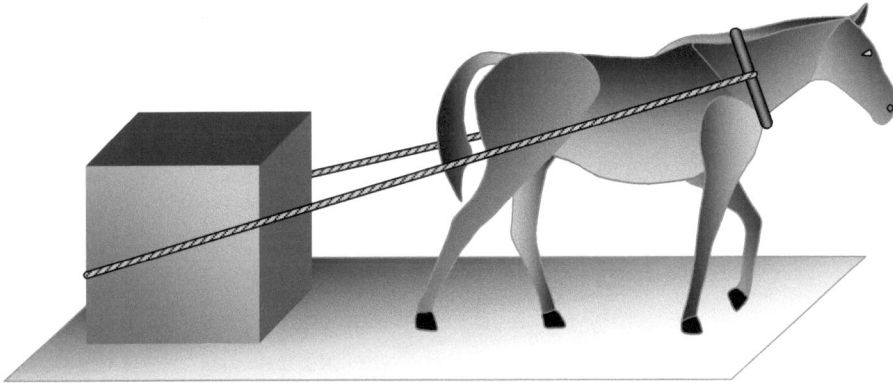

**Figure 4:** My analogy of mechanic work.

The integral of the law of motion also created two theoretical entities known as kinetic and potential energy. Kinetic energy represented a number indicating how fast an object was moving, and potential energy was a number indicating how high this object was in relation to the ground. When everything was put together, scientists found that mechanic work was equal to the change in total kinetic and potential energy. In other words, energy was the mathematical capability of performing mechanic work.

Because of another basic law of physics known as the principle of action and reaction, something that lost energy through performing work was always matched by something that gained the same energy by being the object of this work. In a purely theoretical scenario, energy was never created or destroyed, but only transferred through mechanic work.

## THE PRINCIPLE OF ENERGY CONSERVATION

Even though kinetic and potential energy were indestructible in theory, in practice they were destroyed by friction. Friction was a force that opposed one surface as it moved with respect to another, and because this force was not moving, it had the theoretical effect of making mechanic work disappear.

In the early 1800s, scientists knew that friction generated heat, but they did not have any reason to believe that heat and energy were somehow related. At that time, the most popular theory about heat was the caloric theory (Antoine Lavoisier, 1783), which stated that heat was a material substance, and friction only released the heat that was originally stored inside objects.

There was some evidence, however, that friction could have been what actually created the heat. The first evidence for this was provided by a scientist that generated heat by boring cannons with a blunt drill (Benjamin Thompson, 1798). The conclusive evidence was provided half a century later with an ingenious piece of equipment that was called a calorimeter (James Joule, 1845).

The calorimeter was basically an insulated water container with a set of internal blades. The idea was to wind a cord around the blades' shaft and then allow an external load to unwind it as it fell down. By multiplying the weight of the load by the length of the cord, scientists knew how much work was used to propel the blades, and then by measuring the temperature of the water, they knew how much heat was generated.

The results showed that heat and work were directly related. As far as scientists could know, energy was not created or destroyed, but was transformed from one form into another.

## THE SUBSTANCE OF EVENTS

When heat was translated into energy in the mid-1800s, scientists were no longer able to tell if energy was real or imaginary. Energy started as a pure mathematical definition, but it suddenly escaped from the law of motion and started behaving as if it were real. Regardless of what it really is, energy behaves as a real entity that creates events as it changes and moves.

Events, according to this theory, are just the manifestation of energy moving from one place to another and changing from one form into another. For example, an event that can illustrate this concept is the experiment with the balls at the Leaning Tower of Pisa.

One might say that the experiment started with the sun, when energy from fire was released in the form of light. The light then traveled to the earth and was absorbed by plants, which were eaten by the scientists who digested them to release calories. The calories were used to climb the tower and stored in the potential energy of the balls, and the potential energy was transformed into kinetic energy when the balls were released from the top. In the end, the kinetic energy became heat again when the balls finally hit the ground.

If somebody then would have taken this heat from the ground and traced its transformation back to the source, then he or she would have ended up with a little fire from the sun. The sun could have generated exactly the same heat by warming up the ground with light, but by going through all these transformations, the energy helped complete an experiment that helped humans understand what it was.

## THE EQUIVALENCE OF MASS AND ENERGY

The ideas of mass and energy were both human definitions based on the law of motion, but most people believed they were real because they associated them with concepts like weight and heat.

This belief started to crumble during the late 1800s with the development of the theory of electromagnetism; this theory predicted that light created a force when it reflected on a surface (James Maxwell, 1874). Since a reflection mathematically represented a change in speed, then this force defined a mass according to the law of motion.

The problem faced was that light could be fully transformed into heat, and heat was supposed to be a pure form of energy. In addition, having mass inside light violated the principle of action and reaction. Scientists believed that if light had mass, then a light bulb should be pushed back by inertia, like a cannon that shoots a projectile (Jules Poincaré, 1900).

In the early 1900s, however, the force of light was confirmed experimentally (Ernest Nichols & Gordon Hull, 1903). Scientists measured the force of light reflection by using a very sensitive balance on which the plate that would hold the sample was replaced by a mirror. It became evident that light, which was apparently made of pure energy, behaved as if it had mass by definition.

To solve this paradox, scientists imagined a new form of existence for mass and energy (Albert Einstein, Nov 1905). They assumed that, when a substance emitted light, a tiny fraction of mass dematerialized and propagated through space in the form of energy. Later, this energy materialized into mass again when the light was absorbed by another substance. In other words, mass and energy presented two different forms of measuring the same entity.

The documented equivalence between mass and energy was probably the most famous equation deduced in the special theory of relativity. $\mathbf{E} = \mathbf{mc}^2$, which I personally see as an icon of scientific ingenuity, says that energy is equal to mass multiplied two times by the speed of light.

## THE ATOMIC BOMB

If the special theory of relativity was correct, then mass and energy were two different definitions of the same thing. Things may be perceived as either mass or energy, but in theory they are both mass and energy at the same time. This theory was validated a few decades later through the discovery of nuclear fission (Otto Hahn, 1938).

Nuclear fission was the equivalent of shooting atoms with particles. The impact created microscopic explosions in which the atoms split into pieces and released huge amounts of energy. Scientists found that the mass of the pieces that remained after that explosion was lower than the original mass of the atoms, thus suggesting that part of the mass was dematerialized into energy.

Scientists also deduced that the pieces from these atomic explosions could successively hit other atoms and produce a chain reaction. This was proved to the world in the mid-1900s with the explosion of the atomic bomb (The Manhattan Project, 1941).

## THE THEORY OF THE ONE ELEMENT

If scientists are correct, then the immense variety of objects and events that people perceive merely are representations of the many forms of a single element.

Depending on the observer, everything that happens in the universe can be described through combinations and transformations of mass and energy.

This element is also eternal in the sense that it cannot be created or destroyed. If mass and energy are reduced in one part of the universe, then there must be another part of the universe in which mass and energy increases by the same amount. In other words, the universe is either eternal or was created by something else.

## THE THEORY OF THE BIG BANG

Scientists could only dream about knowing what created the universe until they developed indirect methods to measure the position and speed of the stars.

The position of the starts was basically estimated by measuring the brightness of their light. Scientists assumed the magnitude of the stars' brightness should be at their surfaces, and then calculated the distance by measuring how bright they were as measured by telescopes. In addition, they found also stars that emitted pulses at a regular frequency, and it turned out that the brightness and the frequency of the pulses were directly related (Henrietta Leavitt, 1912).

The speed of the stars was estimated by measuring the color of their light (Vesto Slipher, 1912). Scientists knew that waves emitted by moving objects became narrower to the observer when the object was approaching and wider when the object was receding. For example, when an ambulance passes very quickly by a person, the pitch of the siren suddenly changes from high to low upon observance by that person. In the case of light, the color shifts toward blue when the object is coming and red when the object is going.

By measuring the brightness and color of the stars, scientists estimated the position and speed of the galaxies (Edwin Hubble, 1929). They found that almost all galaxies were moving away from earth, thus giving the clear impression that the universe was expanding. Scientists then proposed competing theories to explain the expansion of the universe, and the one that was finally accepted was the theory of the Big Bang (Monsignor Lemaître, 1931).

According to the Big Bang theory, the entire universe was once concentrated in an extremely small place at an extremely high temperature. It spontaneously started to expand and cool down, and after a few hundred thousand years, mass and energy started to condense and light was released into space. As the universe continued expanding and cooling down, mass gradually condensed into dust and eventually into stars and planets.

## DARK MATTER AND ENERGY

After proposing the theory of the Big Bang, scientists also developed a theoretical method to estimate the mass of the galaxies (Fritz Zwicky, 1933). Simply put, if a galaxy was imagined as a revolving cluster of stars bound together by gravity, then the kinetic energy accumulated in the rotation has to be one half of the potential energy accumulated in the attraction.

The results showed that the mass estimated with this method was more than 100 times greater than the mass estimated with the telescopes. On the basis of this evidence, scientists concluded that most of the mass inside a galaxy was invisible to the senses of perception. They called it dark matter, because it did not emit or absorb light.

With so much normal and dark matter around, it was logical to assume that the expansion of the universe would gradually slow down. If the Big Bang was a huge explosion, then the debris would have expanded very quickly at the beginning and then slowed down gradually by gravity.

In the late 1990s, however, an observed supernova demonstrated the opposite. A supernova is basically a huge explosion of a star, but this one was so far from the earth that the light required billions of years to reach the telescopes. This light gave scientists the opportunity to measure the speed of a star when the universe started to expand, and they found that the expansion of the universe was becoming faster instead of slower with time (Adam Riess, 1998).

Until today, there has not been a conclusive theory to explain why the universe seems to be expanding faster with time, but many scientists believe that it is caused by some form of energy. This energy is supposed to be invisible to human beings, thus it is generically referred to as dark energy.

Of course, the real nature of dark mass and energy remains a mystery for science, but many are positive that dark mass and energy exist because of the measurable effect that they have on the universe. Furthermore, recent measurements with telescopes in space suggested that the universe is made of approximately 95% dark mass and energy, and 5% of the normal type (NASA).

## THE SUPERNATURAL WORLD

Even though mass and energy are just mathematical definitions, they behave as if they were the real substance making all the objects and events. To my surprise, only 5% of the mass and energy available in the universe seem to be visible to the senses of perception.

If the theory of dark mass and energy is correct, then the apparently infinite variety of objects and events are made using only 5% of the substance available in the universe. This, of course, includes human life and intelligence. By simple comparison, it seems reasonable to assume that the other 95% of the substance available in the universe could support a similar variety of invisible objects and events. Again, this could also include supernatural intelligence and life.

A complete skeptic about the existence of supernatural worlds should assume that, while normal mass and energy evolved into an immense variety of things, dark mass and energy remained unchanged since the Big Bang occurred. Besides being an extremely egocentric assumption, this is not even logical from the scientific point view. Dark energy that can accelerate the expansion of the universe and dark mass that can keep the galaxies together obviously have enough forces to generate change.

For me, the discovery of dark mass and energy is almost a scientific validation of spiritual beliefs. Spiritual people have always claimed that there are more elements in the universe than those that are visible to the senses of perception. They have always claimed that those elements are real, and they have the power to affect the order of the universe. Since science has demonstrated that these claims are true, it would be difficult to believe that they are true by coincidence.

As mentioned in the introduction, I once found my friends by just following the idea that I was going to find them. Simple experiences like that fuel the belief so many people hold that there is an invisible connection between the mind and the universe. This text cannot demonstrate that this connection is true, but it can certainly demonstrate that it is scientifically possible.

# CHAPTER 4

## The Age of Time

**Abstract:** "The Age of Time" explores the possibility that the universe may possess memory of the past. The discussion starts by explaining the relationship between time and age. The chapter explains that, even though time is just an imaginary concept, age seems to be something real that irreversibly advances with time. To understand this connection, the chapter quickly reviews the scientific definition of time, from the sundial to the atomic clock, and it then continues with the history of entropy from the steam engine to the theory of communication. It concludes that entropy behaves as an immaterial memory that takes note of every action and remembers it for the rest of eternity. Metaphorically speaking, entropy behaves as if the universe must remember the past in order to avoid repeating it in the future.

**Keywords:** Age and time, old and new, past and future, reversible and irreversible, cycles, ephemerides time, sundial, pendulum clock, atomic clock, perpetual motion machine, the bone digester, steam engine, heat engine, mechanic work, entropy, second law of thermodynamics, kinetic theory of gases, chaos and disorder, theory of communication, universal memory.

## AGING

Perhaps it's just common sense, but people have an intuitive understanding of age. By simply looking at the external appearance of objects, people can guess more or less if these objects are new or old, perceiving common patterns in the appearance of things that are apparently caused by the effect of time.

Of course, one could always say that aging is a man-made illusion. In theory, things may look new or old to the human eye, but this appearance may not necessarily be connected with the scientific measure of time. Hypothetically speaking, something that appeared to be old in the past may change in such a way that it appears to be new in the future. In other words, there is no reason to believe that aging is necessarily caused by time.

Practice, however, shows that there does seem to be a reason. If somebody had invented a material that appears to get newer as it gets older, such a material would be known to everyone. As far as we can tell, age irreversibly increases with time. This may not sound like a revelation for many people, but in the context of this discussion, it clearly suggests that there may be a purpose in aging.

**Figure 1:** A photograph of a new coin and an old coin manufactured in 1951.

## THE PURPOSE OF AGING

The assumption that age is irreversible implies the belief that the future can never be identical to the past. If something goes through a series of transformations and ends being identical to how it started, then these transformations were by definition reversible. An irreversible change must literally create something completely new that has never existed before.

Because everything in the universe is directly or indirectly connected to something else, creating something completely new would necessarily change the balance of the universe. Both the new and the old things would have to find a way to achieve balance again, and this requires making choices and dealing with the consequences. Consequently, the new balance of the universe would not only be determined by the change, but also by the pathway followed toward conditions of equilibrium.

If, for some reason, irreversible change just keeps happening over and over again, then elements that were identical at the moment of creation could evolve through totally different paths, depending on the voluntary or involuntary choices that they made during their existence.

A perfect example of this idea is the Big Bang theory, which states that all objects and events that people perceive today were once the same single concentrated substance at one point in space. Something triggered change in this substance, and it evolved into many things, including those as simple as a pile of dust and those as sophisticated as a human being.

From a scientific perspective, the only difference between a pile of dust and a human being is the cumulative result of their choices made since the moment of creation. Metaphorically speaking, aging forces the universe to seek a purpose in existence. Instead of remaining under equilibrium conditions for the rest of eternity, irreversible changes cause the universe to gradually decide what to become in the future.

A skeptic could always say that a change itself does not need to have any purpose. He or she could always assume that irreversible changes are completely random, thus the consequences of becoming old are only determined by chance and circumstance.

This assumption, however, contradicts the definition of the irreversible nature of the universe. If changes are supposed to be completely random, then there is no reason why a combination of chances and circumstances could not make the future, by coincidence, identical to the past. Belief that age is irreversible implies zero chance to become identical to the past.

## THE CYCLES OF TIME

If aging is really irreversible, then the universe must remember the past in order to avoid repeating it by coincidence in the future. The logical question to ask, therefore, is whether aging is really irreversible or not. Ancient alchemists, for example, believed in a mythological substance called the elixir of life that could cause old people to become young.

Many other people, even some modern scientists, are convinced that the universe may repeat itself in cycles. If the universe exists in cycles, then all possible pathways of evolution should end exactly at the point they started. This is not an unreasonable assumption, since time itself only exists in cycles. Time is not a real element of the universe, but rather a cyclical event that people count over and over again.

For example, the most obvious cycle of time is the duration of day and night. People in the past could have easily measured when a day ended and another day started just by looking at the projection of the shadows on the ground. Furthermore, it is well known that ancient civilizations measured the hours of the day by using an instrument called a sundial.

The sundial can be simply imagined as an obelisk projecting a pointy shadow over the ground. Then, as the sun crosses the sky from east to west, the shadow of the sundial gradually pivots around the obelisk and points to the marks on a graduated scale.

**Figure 2:** My impression of a sundial.

## CALENDARS

Another example of cyclical events available for counting is the seasons of the year. Astronomers were able to know when one year ended and another year started by measuring the distance from the sun to the horizon at noon. Because the days gradually get shorter approaching winter and longer approaching summer, there are two days of the year called solstices on which the sun reaches its closest distance to the poles, and other two midpoints called equinoxes, in which the sun at noon is exactly above peoples' heads.

**Figure 3:** My analogy of using the night sky as a telescope that scan the universe.

For the rest of the year, ancient astronomers developed calendars based on the position of the moon, planets, and stars in the sky at night. The idea was to use the night sky as a gigantic telescope that performed a 360-degree scan of the universe per year. It is believed that astronomers from ancient Babylon divided the universe into 12 zones, each of them marked by a different constellation and defining a time period of approximately one month.

## THE PENDULUM CLOCK

The first mechanical machine to measure time accurately was probably the pendulum clock. Simply described, the pendulum clock runs with a mechanism that can be imagined as a weight hanging from a string.

If one then displaces the weight to one side of the horizontal and releases it from there, then the weight swings all the way to the other side of the horizontal and returns to its initial point. In theory, the weight should return exactly to the same place it was released, thus starting the same cycle again and repeating it for the rest of eternity.

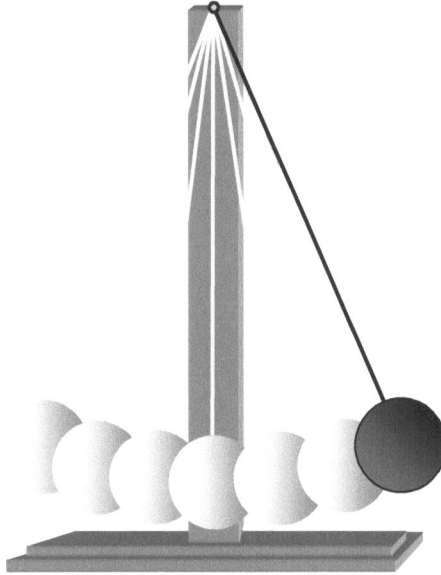

**Figure 4:** My impression of a pendulum swing.

In practice, however, the weight never returns exactly to the same place from which it was released; it always reaches a point closer to the midpoint than the point at which it started, thus the swings get narrower and narrower, and all real pendulums eventually stop.

Fortunately, scientists in the early 1600s discovered that the time required to complete one swing depended only on the length of the string (Galileo Galilei, 1602). This meant that, even though the swings got narrower, the duration of each cycle was always the same.

For its simplicity and accuracy, the pendulum became the perfect tool for measuring time, and by the mid-1600s, the first commercial pendulum clock was designed and patented (Christiaan Huygens, 1656). Since then and until approximately the early 1900s, pendulums were used as the informal reference unit of time.

## THE DEFINITION OF TIME

The only problem with pendulums was how to adjust the length of the string so that all clocks measured exactly the same time. Initially, the pendulum clocks were calibrated by using the cycles of day and night. By the late 1800s, however, it became evident that the cycles of day and night were not really constant. There were random fluctuations in the rotation of the earth around it axis of symmetry that made each day slightly different from the previous one.

Scientists, then, turned to the duration of the year as the reference unit. They knew that the orbit of the earth around the sun was also changing with time, but they were able to calculate this trend by using a mathematical model and ephemeredes data from as far as the mid-1600s (Simon Newcomb, 1895). This mathematical model, which is known today as Ephemeredes time, was used to calculate time based on the position of the sun or the moon in the sky.

A few decades later, however, scientists also invented the atomic clock (Isidor Rabi, 1930); this clock was an electronic device that induced atoms of a pure gas to oscillate in perfect harmony together. This produced extremely stable electromagnetic cycles that could be counted as a measured length of time.

Eventually, scientists forgot about the duration of the year and decided to use the atomic clock as the reference cycle. They defined the second as approximately 10 billion electromagnetic cycles emitted by an element called caesium-133 (General Conference on Weights and Measures, 1967).

## THE PERPETUAL MOTION MACHINE

Measuring time in science involves counting repetitions of a cyclical event; thus there is no reason to assume that time has to make things older. In theory, one could invent a perfectly cyclical device that ends exactly at the point it started so the universe would be identical before and after each a cycle. If this device existed, then measuring time would have absolutely no effect in the age of the universe.

History books record the ever-present dream people have held of inventing the perpetual motion machine. There are known designs of perpetual motion wheels from as early as the 1100s (Bhaskara Achārya, ~1150) to as late as the 1700s (Johann Bessler, 1717). After the 1800s, scientists also looked to electricity and magnetism to unveil the secrets of perpetual motion. A recent example is an electric motor called Perepiteia, which is supposed to generate more power that it consumes (Thane Heins, 2002).

Unfortunately, no perpetual motion machine has ever worked indefinitely. As far as scientists know, each and every perpetual motion machine that people have attempted to make has always stopped at a certain point. Of course, there are those that are not willing to give up the dream, but most scientists today believe that perpetual motion is impossible. The U.S. Patent and Trademark Office, for example, explicitly refuses applications of perpetual motion machines, deeming them useless (USPTO).

## THE STEAM ENGINE

If it were possible to identify the beginning of the end for the perpetual motion dream, that would probably be the bone digester invented in the late 1600s (Denis Papin, 1679). In retrospect, the bone digester was the predecessor of the pressure cooker. It was basically a hermetic metallic container that could be used to boil water inside. Scientists found that, by not allowing the vapor to escape, the container gradually built up pressure and, figuratively, crushed bones.

One problem was that, as the pressure built up, the bone digester would likely explode. Therefore, scientists incorporated a safety valve that released some vapor when the pressure was high. This safety valve was basically a load pressing a piston over a hole. The piston kept the hole closed when the pressure was low, but it allowed some vapor to escape when the pressure was high enough to overcome the weight of the load.

After incorporating the safety valve, the bone digester demonstrated that vapor could push up a load. Scientists immediately realized that a similar mechanism could be used to push the lever of a water pump. By the early 1700s, this idea materialized into the first commercial steam engine used to pump water from the coal mines (Thomas Newcomen, 1712).

The steam engine utilized a cyclical process. First the water was evaporated into steam while pushing a piston forward through a pipe, and then the steam was condensed into water while pulling the piston backward to the initial position.

Each cycle ended exactly as it started, but by pushing a piston back and forth, the engine converted some heat into mechanic work.

## THE DEFINITION OF ENTROPY

After the invention of the steam engine in the early 1700s, scientists faced the first energy conservation problem in history: the engine was first heated up with fire to generate steam and then cooled down with water to condense the steam. Scientists realized that they were wasting too much heat by cooling down the engine after each cycle.

The first improvement was to condense the steam in an external heat exchanger so that the metal did not need to be cooled down after each cycle (James Watt, 1765). This may have reduced heat consumption by 75% and made the steam engine a historic success. Scientists then tried to reduce energy consumption further by optimizing the temperatures that they used to heat up and cool down the engine.

By the early 1800s, however, scientists had mathematical models to describe the relationship among the pressure, volume, and temperature of a gas (Joseph Gay-Lussac, 1802). Therefore, instead of experimenting with a real steam engine, they designed an imaginary heat engine that worked by expanding and contracting a theoretical gas (Nicolas Carnot, 1824).

The heat engine was a perfectly reversible cycle. First the gas was heated up so that it expanded while pushing a piston through a pipe, and then it was cooled down so that it contracted while pulling the piston back to the initial position.

The results showed that it was impossible to convert all of the heat into mechanic work. The gas always had to be cooled down at some point of the cycle, and the amount of heat that had to be released depended on the temperatures. More specifically, they found that the heat divided by the temperature created to expand the gas had to be equal to the heat divided by the temperature released to contract it.

If this conclusion was correct, then the heat divided by the temperature behaved as a substance that flew through the imaginary heat engine and produced a reversible cycle. This imaginary substance was eventually called entropy; scientists in the mid-1600s defined entropy as heat flow divided by its temperature (Rudolf Clausius, 1865).

## THE SECOND LAW OF THERMODYNAMICS

Scientists demonstrated that the imaginary heat engine consumed as much entropy as it released. It remained to be answered, however, whether this was also true for all engines. To answer that question, scientists performed a mental experiment. They mentally connected a real engine to a theoretical heat engine working in reverse.

They then placed the two engines inside a black box, thus they could not see the engines pushing and pulling one to each other, but they could measure the heat coming into and out of the box.

If the two heat engines were identical, then the heat released by one engine would have been completely consumed by the other, and the box would have contained a perpetual motion machine. If, on the contrary, the engines were not identical, then one engine would have released heat from one side of the box and the other taken the same amount of heat from the other side.

By calculating this heat, scientists concluded that the contents of the box could have never consumed more entropy than they released. Otherwise, heat would have crossed the box from cold to hot, and that was supposed to be impossible in practice. This conclusion, which is known today as the second law of thermodynamics (Rudolf Clausius, 1865), says that any spontaneous process in practice always releases more entropy than it consumes.

## THE IMMATERIAL MEMORY

If the second law of thermodynamics is correct, then perpetual motion is impossible. Every time something spontaneously changes or moves, some energy must be degraded into colder heat to increase the entropy of the universe. In the case of the pendulum, for example, the kinetic energy is degraded by friction between the load and the air, thus the swings cannot return to the place they started, and all pendulums eventually stop.

In the context of this book, the second law of thermodynamics also demonstrates that the future can never be identical to the past. Even if the universe for some reason decided to reverse the effect of age, it would never be able to reverse the entropy that the age created. Metaphorically speaking, entropy is like an immaterial memory that always remembers something from each experience of the past.

The characteristic that makes entropy such a mysterious substance is that it does not stay inside the element that created it. The second law of thermodynamics only states that the total entropy of the universe must always increase; thus entropy is free to escape from the setting in which it was created and flow to any other place in the universe. If entropy is again compared to an immaterial memory, then this memory belongs to the universe as a whole.

## THE KINETIC THEORY OF GASES

Since entropy must always increase, the future can never became identical to the past. However, this does not necessarily mean that the universe can remember things. Most scientists would assert that entropy is an amorphous entity that increases for no reason. To believe that the universe has memory of the past, one should interpret entropy as an entity that can store information.

Fortunately, scientists proposed a theory that made interpretation of entropy much more visual. More than 100 years before entropy was even known, scientists proposed the kinetic theory of gases (Daniel Bernoulli, 1738). According to this theory, a gas is made of an extremely large number of molecules flying in all directions and colliding with each other.

Scientists used the kinetic theory of gases to explain heat and temperature on the basis of the laws of physics. They explained that heat was the kinetic energy accumulated in the speed of the molecules, and temperature was the rate at which this kinetic energy was transferred from one molecule to another.

At some point, however, scientists realized that molecules could not be all moving at the same speed; thus they proposed a mathematical equation to calculate the probability of a certain molecule moving at a certain speed (James Maxwell,

1859). This introduced uncertainty in the description of a gas. The equation showed that there are many ways to distribute the speeds among the molecules and get exactly the same heat and temperature.

The solution was to assume that many of these distributions were different versions of others but with the molecules in different orders. This meant that there were many versions of the same gas, and simply by performing random collisions, the molecules tended to the version that provided the maximum number of options.

When scientists calculated the entropy, they found that it was a measure of the number of options that the molecules have inside a gas (Ludwig Boltzmann, 1877). If this interpretation was correct, then entropy always increases, because having more options is the condition that presents more probabilities to occur in the long run.

## THE MESSAGE

Since entropy is a measure of the number of options that the molecules have inside a gas, it has become customary in science to say that entropy is a measure of chaos and disorder. Chaos and disorder have been proven to represent a very graphical metaphor to explain the concept of entropy, but it gives the impression that entropy is a destructive force. In the context of this book, it would be very hard to believe that something associated with chaos and disorder could have a memory of the past.

However, this is not a contradiction at all; the meaning of the message is independent of the elements in which the message is written. A very simple example could be a line of text written in the English language. The letters in a line of text are just the elements in which the message is written, but the message is represented by the order in which these letters appear.

For someone who does not know how to read English, a line of text may look like a sequence of symbols that do not follow any particular order. On the other hand, if a person knows how to read English, then the same sequence of symbols would be an intelligible message. In the same way, the different internal orders of matter may look like chaos and disorder to the human observer, but they may contain intelligible messages for a consciousness that can understand the language.

## THE THEORY OF COMMUNICATION

The apparent chaos and disorder in the order of written letters is just the consequence of having information stored in them. This is because languages need to have a certain degree of uncertainty in order to communicate different messages. If it were always possible to predict the order of the letters before reading them from the text, then language would always communicate the same message.

In any human language, however, there are grammar and spelling rules that make communication a little redundant. For example, a typical grammar test in school asks students to complete a phrase for which some words or letters have been replaced by blank spaces. If the children can pass the test, then it means that the text contains more letters than necessarily to actually understand the message.

In the mid-1900s, scientists realized that having more letters than necessarily was not good for electronic devices; these devices had a limited capability to store and transmit information, thus writing bigger messages made operations slower and more expensive. The solution was to create compression codes that minimize the probability of predicting the message.

The idea was that, if a message were easy to predict, then it contained letters that were not necessary to understand it. Surprisingly, scientists found the chances of predicting a message were mathematically identical to the inverse of the entropy equation (Claude Shannon, 1948). In other words, if the entropy of a message increased, then the message contained more letters with meaningful information.

If the theory of communication is correct, then the equation of entropy can be mathematically interpreted as the amount of information that can be stored in apparent chaos and disorder.

## THE UNIVERSAL MEMORY

Before exploring the more spiritual aspects of science, I'd like to finish this chapter by making a slight but reasonable connection between science and spirituality. Many spiritual people believe that they will one day be accountable for their actions and that they will be judged after death, while others believe that their actions in this life will affect their circumstances in the future.

Even though this connection is slight, science has demonstrated that there is a measurable entity in the universe that could be interpreted as a universal memory. Under this interpretation, the entropy of the universe must always increase, because there are always more things to remember. For the skeptic, the answer is simple; there is an immaterial entity in the universe that makes it impossible to fully reverse the actions of the past.

# CHAPTER 5

## Between Life and No Life

**Abstract:** "Between Life and No Life" explores the question of whether life can exist independent of the physical body. The discussion starts with a brief introduction to the scientific definition of life. It explains that life in science is supposed to exist within microscopic units called cells, and these cells can only be created by dividing other cells in two. The story then talks about the theory of evolution, the motivations for the theory, and how it was apparently validated by the theory of molecular genetics. This chapter concludes that, if the theory of molecular genetics is correct, then there is an obvious paradox in the definition of life; the creation of life contradicts its own definition. Regardless of the efforts to explain this paradox during the last century, scientists have still not found the law of nature that can fill this gap. In contrast, the hypothesis proposed by the author asserts that life behaves as if it is has consciousness of its own existence.

**Keywords:** The wrong assumption, theory of vitalism, organic and inorganic matter, body and consciousness, life definition, capability of self-replication, spontaneous life generation, the theory of evolution, natural selection, common ancestor, inheritance, genes, classic genetics, molecular genetics, chromosome, DNA, mutations, synthetic life, stem cells, sexual reproduction.

### BODY AND MIND

There are psychological reasons to believe that life may transcend the boundaries of the physical body. One reason is that people do not necessarily perceive their thoughts as if they were part of the laws of nature. When scientists think about the laws of nature, for example, they do not feel that the laws of nature involve thinking about themselves. They would rather see themselves as external observers.

A simple mental experiment may help in the visualization of this idea. If one closes the eyes for a moment and avoids thinking about anything that comes from the senses of perception, then the thoughts would perceive themselves as suspended in a dark space. Under this state of pure consciousness, the thoughts can imagine themselves in a different body, but they can never imagine themselves thinking of something different. If life is defined by its own consciousness, then thoughts are the irreducible unit of life.

This belief is also reinforced by the way in which people normally die. For somebody looking at a person about to die, death is the moment in which the person stops having vital signs. Even if the body looks exactly the same before and after death, an observer has the intuitive feeling that life has suddenly gone at that particular moment in time.

## THE THEORY OF VITALISM

In the past it was not uncommon to believe in the existence of vital energy. The ancient Roman theory of the humors, for example, assumed that human emotions were caused by the combination of four basic fluids (Claudius Galenus, ~150). Much more recently in history, the theory of vital energy was referred to as vitalism, and it was part of the scientific debate until at least the early 1900s (Hans Driesch, 1914).

People also previously believed that living organisms were not related to non-living things, traditionally classifying matter as organic and inorganic. This classification, however, started to became obsolete in the early 1800s when scientists created urea in the laboratory (Friedrich Wöhler, 1828). Urea was an organic compound found in the urine of animals, but scientists created it in a laboratory by combining liquids that were supposed to be inorganic.

From that point, scientists became more and more convinced that living organisms were not really different from non-living things. They reached the conclusion that life could be explained by using only the laws of nature. If this interpretation was correct, then living organisms in modern science are the equivalent of very sophisticated automats.

## THE SEPARATION OF BODY AND CONSCIOUSNESS

Most modern scientists believe that living organisms and non-living things are essentially made of the same materials. They believe that it is possible to place the right atoms in the right order, and within the right environmental conditions to create a material that behaves as though it were alive.

This, of course, creates an apparent contradiction between science and spirituality. While spiritual people tend to believe that life is an immaterial consciousness, scientists work under the assumption that consciousness is just a natural phenomenon. The consciousness of life is just the laws of nature thinking about themselves.

This contradiction, in reality, is an issue of definition. While scientists study life as observers of the laws of nature, spiritual people study life as observers of their own consciousness. Depending on the point of view, life could be both a material body and an immaterial consciousness.

There is no guarantee that the scientific definition of life refers exactly to the same concept than life in a spiritual context. Like other scientific concepts, such as the space-time continuum, wave-particle duality, and mass-energy equivalence, the separation of body and consciousness may also be a consequence of the observer.

The problem is that scientific theories always need to be validated by an external observer, which limits the conversation in this book to those aspects of life that can be measured as natural phenomena.

## THE SCIENTIFIC DEFINITION OF LIFE

It seems reasonable in science to define life as a behavior rather than a material. One reason may be that living organisms are made of the same materials than the dead versions of the same organisms. Another reason is that there may be life on other planets, and even in remote areas of our planet, that is made of different materials.

Life according to science is defined as a list of behaviors that are apparently common to all living organisms. If a behavior is called a capability, then this list may include the capabilities of self-replication, adaptability to the environment, ability to heal when injured, consumption of energy resources, and response to external stimuli. The capability of self-replication, however, is the single most important behavior that everybody agrees is included in the definition of life.

## SPONTANEOUS LIFE GENERATION

Defining life as the capability of self-replication is a relatively modern concept. Many people in the past were more inclined to believe that life could also be created through spontaneous generation (Aristotle's Generation, ~350 BC). A classic example of life-spontaneous generation was that of maggots growing from putrid meat; if a dead animal were left to decay on the ground, maggots would eventually emerge from the carcass as if they had grown spontaneously inside.

It was not until the late 1600s that a scientist performed a formal experiment to prove that maggots did not grow through spontaneous generation (Francesco Redi, 1668). The common belief at that time was that maggots would spontaneously grow from organic matter as long as they had enough air to breathe. To prove the opposite, a scientist covered a piece of meat with fine gauze. This prevented flies from laying eggs, thus maggots did not grow on the meat even though there was enough air to breathe.

## LIFE SELF-REPLICATION

Just a few years before the experiment with the maggots took place, a scientist also discovered the first living cells (Robert Hooke, 1665). By looking at plant tissue under the microscope, he saw an internal structure that resembled the cells of a monastery.

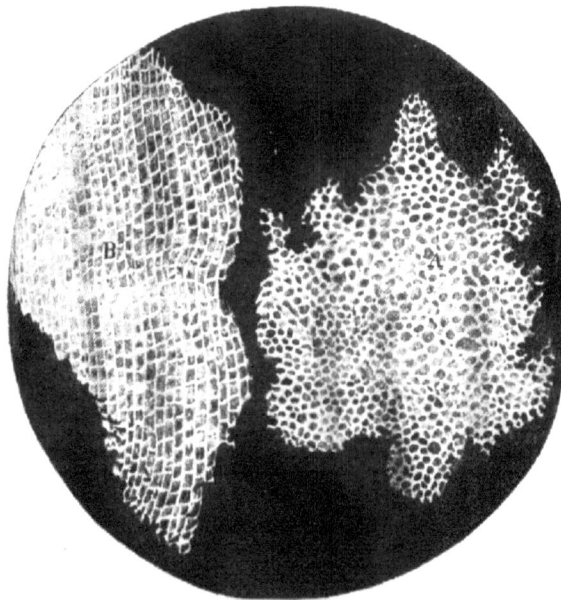

**Figure 1:** Robert Hooke's drawings of plant cells, 1665.

As the technology of the microscope improved, scientists became more and more convinced that cells were independent living organisms. They realized that plant and animal tissue actually consisted of large communities of cells living in perfect harmony together, and microorganisms such as bacteria and some fungi were actually cells living as independent organisms. Almost 200 years after their discovery, scientists finally concluded that cells were the basic units of life (Theodor Schwann & Matthias Schleiden, 1837).

This meant that, regardless of how complex a living organism may seem from the outside, it was always made of smaller units of life. For example, a human being may look like a single individual to another human being, but they are both made of trillions of single individuals living in perfect harmony together.

The theory of cells was concluded in the mid-1800s after scientists explained how cells reproduced. They basically discovered that cells reproduced by dividing their bodies into two (Rudolf Virchow & Robert Remak, 1855). In their own words, they discovered that "*omnis cellula e cellula*", a living cell was always created from a previous living cell.

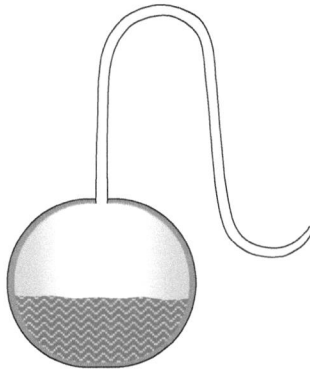

**Figure 2:** My impression of Pasteur's container for testing spontaneous generation.

This, again, contradicted the ancient theory of spontaneous life generation. Some scientists still believed that a microorganism could grow from spontaneous generation if there were enough air to breathe. To put an end to this controversy, a scientist basically repeated the experiment with the maggots but at the microscopic level (Louis Pasteur, 1859). He boiled soup inside a glass container that only allowed air to enter through a long narrow pipe in the shape of an inverted U. Since the microorganisms were too heavy to climb through the vertical section of the inverted U, then only pure air could get inside the container. Of course, the microorganisms did not grow in the soup.

## THE FEATHERLESS CHICKEN

Cell self-replication is such a unique process in nature that it is used almost as a synonym for life. When scientists want to define life in the simplest way possible, they say that it is the capability of self-replication. This definition, however, does not necessarily contradict the spiritual interpretation of life. The capability of self-replication is not life itself, but a measurable behavior of something that is apparently alive.

A simple analogy can be found in the definition of man proposed by ancient Greek philosophers in the 5th century BC (Socrates, ~450 BC). Ancient Greek philosophers reached the conclusion that only humans and birds had the capability to walk on two legs. However, since only birds seemed to have feathers, they agreed to define man as a featherless biped.

The problem was that a featherless biped was not a man, but the appearance of something that was supposed to be a man. To prove this point, another philosopher pulled the feathers from a chicken and introduced it to his colleagues as a man by definition (Diogenes the Cynic, ~350 BC). Philosophers then admitted that the definition was incomplete, and they agreed to redefine men as featherless bipeds with flat nails (Plato, ~350 BC).

The exact same practice applies to the definition of life. In fact, scientists found the equivalent of the featherless chicken while studying a terminal illness that is popularly known as mad cow disease. The disease behaved more or less like a virus in the fact that it killed animals and humans by replicating in their brains, but unlike viruses, the infectious agent of mad cow disease could not be killed with applied radiation.

The reason for this was that the infectious agent was a protein (Stanley Prusiner, 1982). Proteins are one of the most important components of life on earth, but a protein alone is not considered alive even if it has the capability of self-replication.

## THE THEORY OF EVOLUTION

It was perhaps a coincidence that the death of spontaneous life generation (Louis Pasteur, 1859) coincides with the birth of the theory of evolution (Charles Darwin, 1859). Before that year, many people believed that species were immutable (Jean Cuvier, 1830). They claimed, for example, that animals and people mummified thousands of years ago did not look any different from those in modern times.

This was also consistent with religious beliefs. Some religions conceptualized species as designs, while individuals were just good or bad examples of those designs.

Many of these people, however, also understood the concept of inheritance. It is reasonable to believe that everybody would have expected to see a child who resembled his father. On the other hand, seeing a child who resembled another man would have been much more difficult to explain. In general, it should have been relatively obvious that descendants inherited their appearances from their progenitors.

A visual analogy to illustrate the theory of evolution was the technique of selective breeding. In simple words, selective breeding consisted of letting only the best animals and plants reproduce and pass their hereditary advantages to the next generation. For example, a farmer could have decided that the best cows for reproduction were those that produced more milk, while the best horses for reproduction were those that run faster than the rest.

The theory of evolution was nature's version of selective breeding. The idea was that animals and plants had to compete for the same resources and survive under the same environmental conditions. Consequently, only those that were better adapted to survive and reproduce were able to pass their hereditary advantages to their descendants.

The difference was obviously the observer. In natural selection, the hereditary advantage has to spread from generation to generation until it is finally assumed by the entire species. By the time it is fully incorporated into the species, it is not an advantage anymore, thus evolution was virtually invisible to the human eye.

## THE MONKEY MAN

Since evolution was imperceptible, scientists had to rely on assumptions to provide experimental validation. The idea was to assume that species that look different today may have evolved from the same common ancestor in the past. If this was true, then similarities among the species would have provided evidence of a common ancestor, and differences among the species would have provided evidence of evolutionary changes.

Scientists then compared animals and plants that lived in different parts of the planet and found similarities consistent with the theory of the common ancestor. They also compared the fossilized remains of animals and plants that lived thousands or millions of years ago. By putting those fossils in chronological order, scientists created the illusion of a gradual transmutation from one species into another.

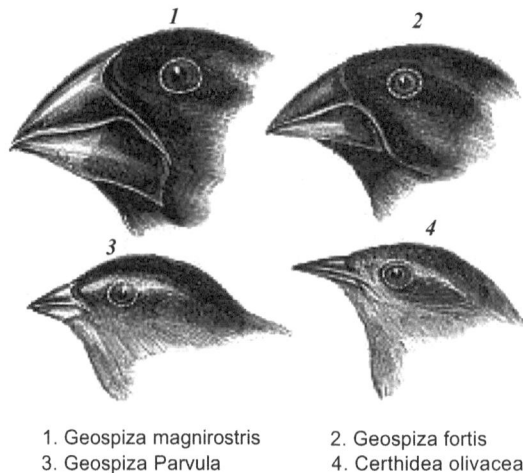

1. Geospiza magnirostris     2. Geospiza fortis
3. Geospiza Parvula          4. Certhidea olivacea

Finches from Galapagos Archipelago

**Figure 3:** Darwin's finches supporting the theory of natural selection, 1845.

The evidence, however, was circumstantial. Scientists did not prove that similar animals or fossilized skeletons in chronological order were actually members of the same family tree; they only proved that it was possible. Even to the trained eye, it was very difficult to find hereditary trails by just comparing creatures that looked more or less alike.

A good example of such circumstantial evidence was a skull found in the early 1900s that seemed to be that of the common ancestor of monkey and man (Piltdown Man, 1912). It took scientists more than 40 years to finally figure out that this apparent missing link between monkey and man was actually a hoax. It was made by combining a human skull, an orangutan jaw, and chimpanzee teeth.

**Figure 4:** My impression of the hypothetic missing link between monkey and man.

## CLASSIC GENETICS

Another flaw in the theory of evolution, at first, was that it did not explain the mechanism of inheritance. It was simply unknown which advantages or disadvantages were actually passed from progenitors to descendants. The theory of acquired characteristics, for example, assumed that the descendants inherited the skills and muscular complexion from their progenitors during life (Jean Lamarck, 1809).

It is known today that the correct mechanism of inheritance was first discovered by a monk in the mid-1800s (Gregor Mendel, 1865). He fertilized pea plants with the pollen of another variety of pea plants and then traced how different features were passed from generation to generation.

His results suggested that information was passed in discrete units of inheritance that were later called genes. Each plant had apparently two copies of the same gene – one called dominant and the other called recessive. The dominant copy determined the external appearance of the plants, while the recessive copy remained hidden inside.

During reproduction, the progenitors passed only one copy of each gene to the descendants. The descendants received two copies, one from the mother and one from the father, and either copy could become dominant regardless of whether it was dominant or not before. One textbook example is a child who inherits the eye color of a grandparent instead of a parent. This shows that the grandparent's eye color was recessive in the genes of one parent and then resurfaced as dominant in the genes of the child.

This study with the pea plants, which is known today as the theory of classic genetics, was rediscovered in the late 1800s and validated independently (Hugo de Vries & Carl Correns, 1900).

## THE CHROMOSOME MOLECULE

In classic genetics, the genes were imaginary entities, but scientists were also convinced that they were made of something real. The best candidate was a molecule found inside the cells a few decades earlier (Friedrich Miescher, 1869). By using a special dye, scientists discovered that cells always received these molecules from a previous cell (Walther Flemming, 1882). These molecules were called chromosomes in reference to the word *color*.

Scientists then connected the imaginary genes with the real chromosome molecules. They showed that, consistent with the rules of classic genetics, normal cells had two copies of each chromosome molecule, but sperm and ovum cells only carried one copy for reproduction (Walter Sutton & Theodor Boveri, 1902).

The conclusive connection, however, was made by a scientist who tested the theory of evolution in the laboratory (Thomas Morgan, 1910). He wanted to know if it was possible to create a new species of flies by irradiating their chromosomes with an X-ray, for example. He apparently could not create any new species, but he created a mutant fly with white eyes. He then demonstrated that new generations of flies inherited the white-eye chromosome as predicted by the theory of classic genetics.

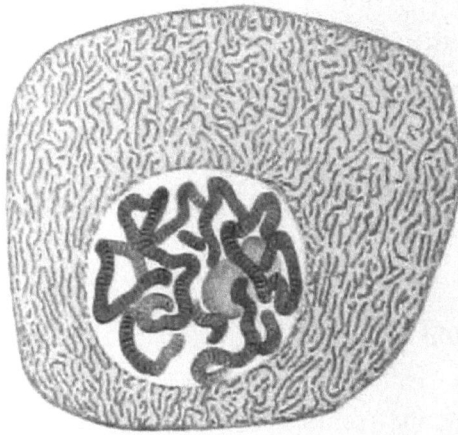

**Figure 5:** Walther Flemming's micrography of chromosomes inside a cell, 1882.

## MOLECULAR GENETICS

By the mid-1900s, scientists had all the information they needed to propose the theory of molecular genetics (James Watson & Francis Crick, 1953), which explained that chromosomes were made of DNA.

DNA was visualized as an extremely long sequence of molecules called base pairs. There were only two different base pairs, and each base pair was in turn made of two different molecules. One base pair was made of an "A" molecule connected to a "T" molecule, and the other base pair was made of a "G" molecule connected to a "C" molecule.

Then, the theory explained that the order of these A-T and G-C base pairs along the DNA sequence encoded genetic information. If genetic code were compared to a computer program, then the DNA molecule would be the equivalent of a long sequence of zeros and ones.

Of course, nobody knows what those zeros and ones really mean, but just by looking at the DNA sequence, scientists can identify and study genes. That is, a gene is the section of the DNA that is found only in organisms with a particular genetic feature.

## DNA REPLICATION

The theory of DNA also explained the process of life self-replication. A simple way to visualize this process is to imagine the DNA sequence as a very long staircase in which the A-T and G-C base pairs are the steps. Replication consists of cutting each base pair in the middle so that one molecule is displaced to the left and the other molecule is displaced to the right.

Immediately after cutting, a new molecule equal to the one on the right is attached to the one on the left, and a new molecule equal to the one on the left is attached to the one on the right. The result is two identical stairs, each having one side completely new and one side coming from the original staircase.

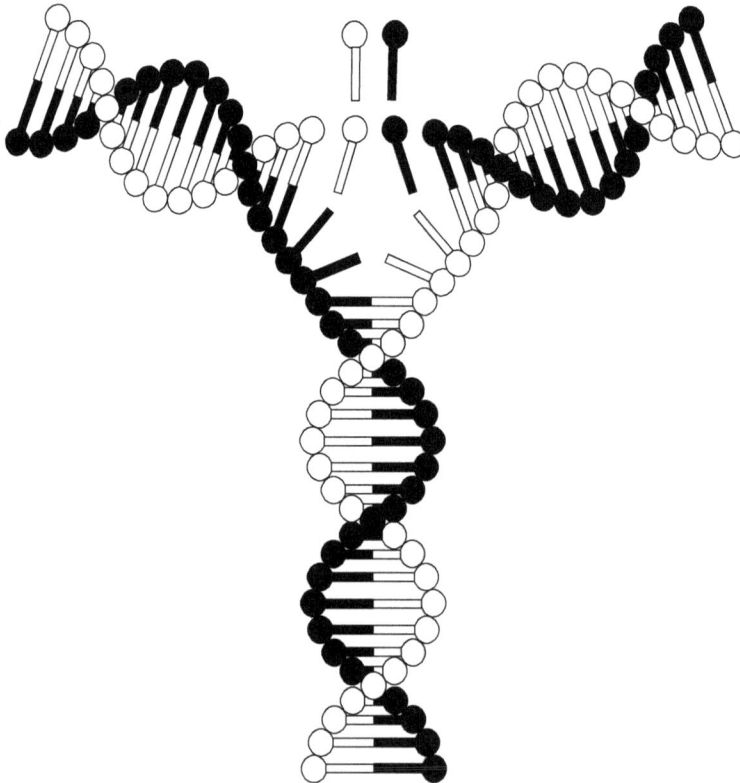

**Figure 6:** My interpretation of DNA self-replication.

## THE COMMON ANCESTOR

If the process of self-replication were perfect, then each descendant DNA should be identical to its ancestor. Furthermore, each DNA molecule should be both half progenitor and half ancestor. Scientists believe that evolution is possible because the cells may create inconsistencies when duplicating the DNA. These inconsistencies are called mutations, and they may lead to self-duplication into a new form of life.

Since mutations only change a small section of the DNA molecule, individuals with mutated genes still inherit most of their genes from previous generations. This means that, if different living organisms have identical genes in common, then they probably inherited those genes from the same common ancestor. By using this rationale, scientists found reasons to believe that all forms of life evolved from the same common ancestor (Douglas Theobald, 2010).

## THE LIVING PARADOX

Of course, the first common ancestor could not have been created from previous life. The first common ancestor had to be created before life even existed. This in turn proposes an obvious paradox in the scientific definition of life. By assuming that life can only be created through self-replication, one necessarily assumes that life was created by spontaneous generation.

Nobody really knows how this happened, but most scientists assume that a fortuitous event a few billions of years ago put the right atoms in the right order and in the right environmental conditions to create the first self-replicating cell. For purposes of the scientific debate, this fortuitous event has to be caused by the laws of nature, but believing in the laws of nature caused the paradox in the first place.

If life was created by the laws of nature, then it is reasonable to assume that the laws of nature have the capability to create life. However, scientific evidence suggests that spontaneous life generation is impossible in nature. There is obviously a missing link between the facts and theory making spontaneous generation and self-replication interdependent, but yet incompatible.

## THE MIRACULOUS EXPLANATION

Some scientists believe that spontaneous life generation only happened once, since it is such an extremely unlikely event. When scientists calculated the odds of having a few hundred molecules randomly combining into a DNA molecule, they simply found that the odds were too low to happen again (Carl Sagan, 1973). In theory, the probability of spontaneous life generation could be so low that it would behave more like miracle than a law of nature.

The problem is that most scientists do not often believe in miraculous events. They prefer to work with assumptions that are measurable and replicable. In fact, most modern-day scientists would believe that spontaneous life generation also occurred on other planets. There is such certainty that we are not alone in the universe that every time somebody finds water on another planet, scientists present it to the general public as the possibility of the existence of extraterrestrial life.

## THE GENESIS OF LIFE

Since the 1950s, scientists have been looking for that legendary law of nature that enabled spontaneous life generation once. The first experiment simply involved recreating the environmental conditions that were supposed to exist at the moment of life creation (Stanley Miller & Harold Urey, 1953).

The idea was to put water and the chemical equivalent of the early earth atmosphere inside a glass container, and then simulate a massive thunderstorm by generating vapor and applying electric discharges. After approximately one week of operation, the thunderstorms generated a large variety of organic compounds, but none of them was, by definition, alive.

Scientists then speculated about other environmental conditions that could have been appropriate for spontaneous life generation. They found potential candidates in the bottom of the oceans, below the surface of the earth, close to deposits of radioactive material, and in meteorites that could have collided with earth. Some more radical scientists even believed that life could be created from chemical reactions in which the molecules catalyze the production of themselves (Stuart Kauffman, 1993).

Just recently, scientists found an unconventional form of life in a U.S. lake that contains too much arsenic for normal life to survive (Felisa Wolfe-Simon *et al.*, 2011). Nevertheless, despite all of these claims, nobody has ever shown any solid evidence of spontaneous life generation other than the theory of the common ancestor.

## SYNTHETIC LIFE

The closest scientists have come to creating life in a laboratory was the manufacture of synthetic life in the early 2000s (Craig Venter, 2007). The creators claimed that they were able to manufacture a DNA molecule in the laboratory and then insert it into a living cell. Once the cell received the synthetic DNA, it started to self-replicate as a new form of life by definition.

Because the cell was alive when it received the new DNA, synthetic life did not create the capability of self-replication. Synthetic life essentially demonstrated that the cells work more or less like a little computer. The analogy consisted of imagining the DNA molecule as the software containing genetic code and the body of the cell as the hardware that reads and executes commands. In the words of its creator, "*the chromosome in the cell is the software*" of life.

## VITALISM REVISITED

If cells can be imagined as little computers that read DNA code and execute commands, then there is still flexibility in the definition of life to accommodate an immaterial consciousness. Like in any computer, there must be a reason the hardware reads the software and executes the commands. There would be no reason to build a computer if no one would use it.

Of course, a skeptic person could always claim that the universe does not need a reason to create a computer. In theory, the universe could repeatedly toss the components up in the air, and they would eventually fall down into place as a fully operational computer. The problem is that this seems impossible in practice. As many times as scientists have thrown the necessary components up into the air, they never fall down as a self-replicating cell.

By using science alone, the scientific definition of life contradicts its own creation. The laws of nature cannot easily explain why spontaneous generation stopped when self-replication started. On the contrary, if one believes that a user created a computer from scratch, then he or she may just have decided to improve it rather than create it again.

Again, the skeptic may say that adding a spiritual dimension to the definition of life is contrary to scientific facts. He or she could argue that this was already rejected in the theory of vitalism. Whether a spiritual person wants to call it spirit, vital energy, or consciousness of life, science has conclusively proven that it does not exist.

However, here I exercise the assumption proposed earlier in this book: The wrong assumption. As explained previously, the scientific method cannot validate invisible entities but only the measurable and predictable effects of these entities on the natural world. If the entity is an invisible consciousness, then its decisions may not be predictable for the human observer. The spiritual dimension of life could be invisible to the scientific method.

## THE IDEA OF THE BODY

Most spiritual people probably believe that life exists independently of the physical body. This may sound like a very unscientific assumption, but it states exactly what it apparently happens when many cells get together to create multi-cellular organisms. In a multi-cellular organism, the idea of body seems to exist even if the body does not.

The most compelling evidence is the behavior of stem cells in animals. Modern biology explains that animals, at the moment of conception, consist of only one single cell. This cell immediately starts creating identical replicates called stem cells, and after some days after conception, the animal becomes an amorphous mass of stem cells known as an embryo.

At some point, however, something almost unimaginable happens. The stem cells realize that they are going to be part of a body and change their physical appearance to materialize that idea. Depending on their relative position inside the mass of cells, the stem cells each change into one of hundreds of different types of cells. For example, stem cells that realize they are going to make up the brain change themselves to process electricity, while stem cells that realize they are going to be the bones change their makeup to accumulate calcium.

After the body of the animal is fully formed, some stem cells remain available for maintenance. If the body is hurt, for example, the stem cells naturally migrate to the wounds and regenerate the tissue. One of the most extreme examples is observed in a water worm called planaria. If a planaria is cut in half, for example, the stem cells in the head will create a new body, and the stem cells in the body will create a new head.

To the external observer, it appears as if the stem cells visualize the idea of a body before the body actually exists.

## ANOTHER PERSONAL STORY

Modern science has conclusively proven that the idea of the body is written in one's genetic code. However, each cell contains something, natural or supernatural, capable of realizing this idea and materializing itself and its neighbors into the physical body. Somehow, the cells must create a body that does not even known that they exist.

The skeptic could argue, however, that the cells do not need a consciousness to create the idea of a body. If the cells are the equivalent of little computers, then a body could be imagined as a huge Internet connection operated from a centralized brain. The cells would be like the nuts and bolts of a very sophisticated automat that just follows commands based on an unconscious algorithm.

A counterargument for this is that many living organisms do not even have a brain, and their cells are still connected through the common idea of a body. I personally learned that simple lesson when I was very young.

One day I decided to decorate a small tree that grew in my family's backyard with festive Christmas baubles. Most of the decorations were just colored glass balls hanging from the ends of the branches, but the most important decoration was a pointy star that had to be placed on the topmost vertical twig.

Because the base of the star was relatively small, I decided to take a knife and cut the needles out from the twig. This provided the perfect conditions for the decorations, but it turned out to be very bad for the tree. Soon the top twig started to turn brown, and it eventually died.

After looking to what I did I thought I had killed the top section of the trunk. Without that portion of the trunk, I thought, the tree would not be able to grow any taller. The following spring, however, something amazing happened: one of the horizontal twigs that had grown to become a branch decided to change course and transform itself into a trunk.

From my perspective, any of the three horizontal twigs close to the top of the tree could have become the trunk. However, the twings never competed with each other; one sole twig completely changed course while the others stayed where they were as if nothing had happened. In my mind, it was obvious that the tree had made an intelligent decision, and all parts of the tree knew about it.

Of course, there is always a reasonable scientific explanation for everything, but one should not automatically discard the spiritual possibility that plants can understand their bodies even though they do not have a brain to think about it.

## LET'S TALK ABOUT SEX

Even the theory of evolution would be easier to understand if we knew that cells had an unrealized consciousness. According to the laws of natural selection, evolution favors those who multiply faster than the rest, but today most living organisms had evolved to have sex. Sexual reproduction consists of blending two cells into one; thus scientists believe that this should be slower than dividing one cell into two (Maynard Smith, 1978).

The process of dividing one cell into two, however, does not offer the possibility of that cell to decide how to evolve. Mutations happen by chance, and natural selection then decides which mutations can stay and which mutations should go. Sexual reproduction, on the other hand, requires the agreement between a male and a female, thus progenitors can select the hereditary features that they would like to see in their descendants.

If cells have an unrealized consciousness, then sex enables them to evolve toward their own ideal of perfection.

## CHAPTER 6

# The Inner Truth

**Abstract:** "The Inner Truth" explores the concept of spirituality from the perspective of the human mind. The chapter starts with a basic introduction to the theory of neurons and neurotransmitters and then explains how they work by using the theory of artificial neural networks as a mathematical analogy. It reveals that, if the theory of artificial neural networks is an acceptable model of the brain, then the human mind cannot arbitrarily create any idea that it wants to; ideas are apparently created from preconceived thinking patterns that were learned by trial and error in the past. The discussion then provides a quick introduction to statistics. It explains that statistics can be used as a tool to measure ambiguous human concepts such as words and feelings. It concludes that, if the theory of evolution is correct, then the brain may contain a spiritual inner truth that explains why so many scientists today do not see any conflicts between science and spirituality.

**Keywords:** Ideas, thoughts, brain cells, neurons, the neuron doctrine, neurotransmitters, electrochemical brain activity, artificial neural networks, neural network training, artificial intelligence, instincts, spiritual beliefs, game of chance, statistics, probability distribution, random error, real value, representative sample, indoctrination, the inner truth.

## THE MASS DELUSION

Many people have spiritual beliefs that are not consistent with science. Common sense would suggest that nothing can be true if it contradicts experimental facts, but some people may not even accept facts if they contradict their spiritual beliefs. Spiritual beliefs are sometimes so irrational that some scientists have deemed them delusion (Richard Dawkins, 2006), claiming that spiritual beliefs are false ideas that persist in spite of contradictory facts.

The obvious fallacy in this statement is that facts are also mere ideas in the mind of the observer. As mentioned before in this book, the only tangible facts that people perceive are light through the eyes, sound through the ears, temperature and pressure through the skin, and certain chemical elements detected through the tongue and nose. In the mind of the observer, any ideas can be true as long as they do not contradict these few facts.

A concrete example is the idea of color. Colors may look absolutely real to the human observer, but they do not actually exist. An observer of the absolute truth, in theory, should look at color and see electromagnetic waves. Furthermore, some ideas are considered true even though they contradict the senses of perception. Scientific concepts such as the space-time continuum, wave-particle duality, and mass-energy equivalence are clearly contradictory to common sense.

Saying that spiritual beliefs are false ideas that contradict the facts is like saying that a colorblind person is delusional because he or she cannot see color. It is very hard to judge another person's beliefs as irrational without being able to see them as they appear in that person's mind. Without being able to see what he or she sees, a delusional person might just be one who sees the world in different colors.

Of course, the skeptic could argue that if only one person sees the world in different colors, then the rest of the people could assume that he or she is the delusional one. The problem with this argument is that, in many cases half of the people see the world in one way and the other half sees the world in another. If everybody believes that they are the observers of the absolute truth, then each half would believe that the other half is delusional. There is no simple way to resolve this disagreement since everybody is merely looking into their own minds.

## THE NEURON DOCTRINE

While one may not judge another's ideas without seeing the ideas in his or her mind, there is little to learn by just looking at a human brain; one would only see a mass of grey tissue with the consistency of gelatin.

The first clue for understanding the human brain came from cutting frogs and connecting them to electricity. The same scientist in the late 1700s who noticed that the legs of a dead frog would move again when he accidentally touched its spinal cord with electricity (Luigi Galvani, 1791), made it obvious that the movement of animals' muscles was controlled by electricity.

In the mid-1800s, scientists also discovered that all living tissues were made of cells (Theodore Schwann & Matthias Schleiden, 1838). However, brain tissue seemed to be the only exception. Under a microscope, brain tissue looked like a homogenous substance made of entangled fibers. It was not until the late 1800s that scientists were able to see the first brain cells by using a staining technique known as black reaction (Camillo Golgi, 1873).

This staining technique facilitated a chemical reaction that colored a few brain cells black while leaving the rest of the brain tissue untouched. By looking at the original drawings, these stained cells looked more or less like a deciduous tree in winter. They seemed to protrude into a thick trunk that gradually subdivided into thinner and thinner branches.

On the basis of these observations, scientists came up with the electric theory of the brain (Santiago Ramón y Cajal, 1891). They assumed that brain cells were like little electricity generators, and those braches they saw as a result of the black reaction were the equivalent of electric cables. If the brain cells were again compared to a deciduous tree, then the electricity was supposed to be released into the trunk and distributed to other parts of the brain through the branches.

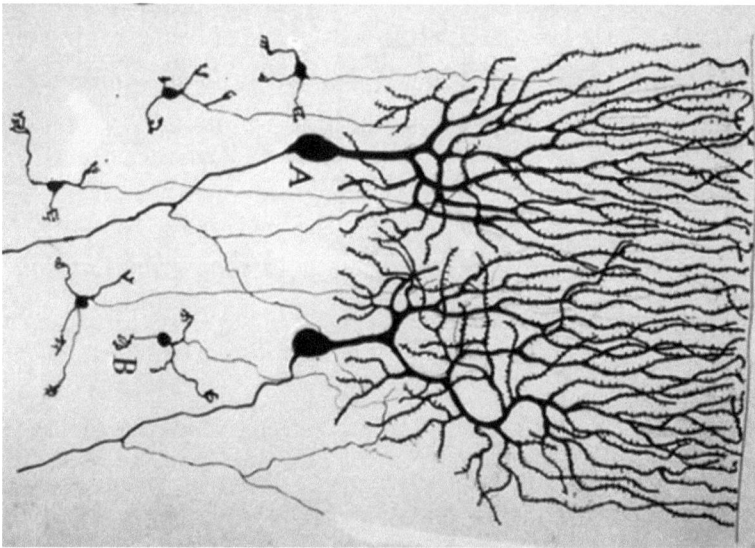

**Figure 1:** Santiago Ramón y Cajal's drawing of brain cells, 1899.

## NEUROTRANSMITTERS

By the early 1900s, the brain cells had been named neurons, and scientists were convinced that these neurons communicated using electricity. The only problem was that the scientists did not see these electric connections using the black reaction. Under the microscope, the black reaction showed tiny gaps between neuron connections that were not consistent with the idea of electrical contact. It was evident that electricity could not flow directly from one neuron to another in such a setting.

The answer to that question provided the final clue for understanding the human brain, and curiously, it also came from cutting frogs and connecting them to electricity. Like in the stories of pre-Columbian Mayan sacrifices (Diego de Landa, 1566), scientists knew than one could cut out the heart from a living animal and it would continue beating for a while at a normal rate. In addition, scientists also knew that they could change the heartbeats of a freshly cut heart by connecting it to electricity.

One day, one scientist had a dream (Otto Loewi, 1921). He cut out the hearts of two frogs and connected one to electricity. Then he extracted fluids from the heart connected to electricity and injected them into the other heart. The result was that the two hearts ended up sustaining similar heart rates, thus suggesting that they were controlled by a chemical that the first heart released into the fluids it pumped.

In other words, the neurons send electricity to other parts of the tissue, but they convert it into chemicals that transfer the electricity to other neurons. Since that experiment, scientists have discovered many more of these chemicals that they called neurotransmitters; they found excitatory neurotransmitters that made the connection easier and inhibitory neurotransmitters that made the connection more difficult to achieve. Ultimately, the communication between neurons depended on the balance of neurotransmitters released into the gaps.

## READING THE MIND

If the theory of neurons is correct, then ideas are basically made of electricity and neurotransmitters. It should be possible in theory to read a person's mind by measuring the electrochemical activity inside his or her brain.

To a certain extent, scientists have developed technologies that can literally read the human mind. For example, electro-encephalography (Hans Berger, 1924) and magneto-encephalography (David Cohen, 1968) are techniques to measure the electromagnetic waves generated by the brain. Unfortunately, these techniques can only measure the combined activity of many neurons; they can certainly show which neurons are thinking, but they cannot really tell what they are thinking about.

## ARTIFICIAL NEURAL NETWORKS

Mathematicians in the mid-1900s proposed a much more theoretical approach to explain human thoughts. They proposed mathematical models that simulated the way the human brain was believed to generate ideas (Walter Pitts & Warren McCulloch, 1943).

First, they created artificial neurons that were basically mathematical equations that had many inputs and only one output. For example, the simplest neuron was an equation that added all of the input numbers and generated an output only if the sum was higher than a certain threshold value. Then they created artificial neural networks by connecting the outputs of many artificial neurons to the inputs of many others.

To simulate the effect of neurotransmitters, scientists created the multiplier numbers. A multiplier was just a constant number that multiplied one output before it was supplied to other neurons as an input. A multiplier higher than one simulated an excitatory neurotransmitter, and a multiplier lower than one simulated an inhibitory one.

The simulation resulted in a mathematical arrangement that, in theory, created ideas out of numbers. Ideas were numeric patterns distributed over an internal area of neurons, while the problems and the results of thinking were represented by numbers exchanged with the external world.

## LEARNING HOW TO THINK

When artificial neural networks were first created, they did not really think by themselves. They were designed to think what the programmer wanted them to think. This is because the programmer had to design the neuron functions, connect the inputs and outputs, and then set up the multiplier numbers.

A human brain, on the other hand, has the capability to solve new problems. If it fails to solve a problem the first time, then it could still generate new ideas and find solutions later. It was clear that scientists needed to generate new ideas if they really wanted to simulate the human brain.

The solution was to propose a mathematical model for learning. Scientists assumed that neurons choose their neurotransmitters depending on how often they needed to be used (Donald Hebb, 1949). For example, if a neuron is used too much, then it probably releases more excitatory neurotransmitters, while if a neuron is barely needed, then it probably releases more inhibitory ones. In other words, learning was simulated by enabling the artificial neural network to adjust its multiplier numbers.

## LEARNING HOW TO READ

In order to teach an artificial neural network how to think, it was exposed to problems for which the solutions were known and then asked to change its multiplier numbers so that it ended up producing the same solutions. A good example of learning in artificial neural networks is a piece of software that can be trained to read the owner's handwriting.

Learning how to read is not a trivial problem, because the figures representing letters are not necessarily identical every time someone writes them. The reader needs not only to see the geometry of the figures, but also the symbolism that was originally present in the writer's mind. That is, the reader must learn how to think like the writer.

The learning process in this piece of software starts with instruction to the owner to write a letter of the alphabet. Depending on the network design and initial multiplier values, the handwritten figure generates a random numerical pattern throughout an area of neurons.

After that, the software asks the owner to write the same letter again. The new handwritten figure would be slightly different from the first one, thus it generates a different numerical pattern inside the network. This time, however, the software adjusts the multiplier numbers to generate one common numerical pattern representing the two figures.

After repeating the same procedure over and over again, the software associates one common numerical pattern to many versions of the same letter, and at that point it decides that it has already learned. At that point, the software has learned how to think like the owner, thus becoming able to read any new handwritten letter by drawing upon only previous experience.

## ARTIFICIAL INTELLIGENCE

The process of training enables a neural network to learn how to solve problems for which the solutions are known, and then the neural network uses this experience to solve new problems for which the solutions are unknown. If intelligence is defined as the capability to solve new problems, then a trained artificial neural network is a form of intelligence by definition.

In fact, teaching a neural network how to read is not really different from teaching a child to do the same thing. The task of learning to read basically consists of repeating pronunciations that are associated to the letters. One may hear many different pronunciations of the same word until he or she is able to create a physiological connection between the sounds and the letters.

Even though artificial neural networks are infinitely less intelligent than human beings, I am not aware of any scientific study suggesting that the theory is essentially wrong.

## THE EVOLUTION OF IDEAS

If one assumes that artificial neural networks are a reasonable interpretation of the human brain, then one also assumes that the brain cannot create any idea that it wants. Theory states that the brain creates ideas by interconnecting neurons and training connections by trial and error.

This is not very difficult to believe, since everybody knows that people need practice in order to master a new skill. Even the senses of perception need training to create ideas. If something blocks the eyes of a child during the first 6 years of life, for example, then he or she may never be able to see, even if the blockage is removed later.

Other basic human skills, on the other hand, must be fully operational immediately after birth. The brain of a newborn baby must be able to maintain all vital functions of the body as soon as he or she leaves the womb. Furthermore, the baby must immediately recognize his or her mother and seek food and protection. It is obvious that a baby is born with a fully operational brain that was trained by somebody else.

If the theory of evolution is correct, then this should not be difficult to explain. One could claim that the evolution of the brain was no different from the evolution of any other human organ. That is, one could just assume that some neuron connections are hereditary and that they have been trained through natural selection.

## INSTINCTS

The existence of instincts provides clear evidence that some ideas are hereditary. For example, most people are intuitively afraid of rats, spiders, and snakes immediately upon seeing them for the first time. It is clear that these individuals are not afraid of the creatures because they have been previously hurt by one; they are afraid of them, in theory, because their ancestors were killed by rats, spiders, and snakes before they could pass any idea to the contrary onto the next generation.

Today, an educated person is able to assess the risk and decide if these creatures are a threat or not in a particular situation, but the instinctive feeling of rejection would still be there, because evolution does not have a mechanism to erase obsolete information. If a hereditary idea does not reduce the chances of survival and reproduction, then it remains in the collective mind of humans for many generations to come.

Metaphorically speaking, the human mind contains a library of ancient ideas that was inherited from the minds of our ancestors.

Having instincts today can be compared to having toes on the feet. Toes seem to be as sophisticated as fingers, but we only used them to walk; they probably tell us that our previous ancestors may have needed hands in their feet in order to climb the trees. In the same way, some instincts may sound irrational today, but they are talking about circumstances in the past for which those ideas were probably reasonable.

## THE IMAGINARY WINDOWS

When a person thinks of something by instinct, he or she may be looking at the vestiges of a reality that was experienced by somebody else. Again, as a metaphor, he or she may open imaginary windows into the ideas of a previous life.

In many cases, ideas that were real in the minds of others in the past may not look real to the human eye, but they are not disputed by science because they can be easily interpreted in the context of evolution. For example, most people would find it almost impossible to explain the idea of love, but it is easy to explain how this idea could have increased the chances of survival and − especially − reproduction.

Spiritual beliefs, on the other hand, do not have any scientific explanation, thus skeptic people assume that these ideas are false. However, the fact that people cannot understand what they see does not necessarily mean that what they see is false. It is a fallacy to say that everything that is real inside the human mind has to look real to the human eye.

## THE ILLUSION OF COMMUNICATION

Nobody who believes in the theory of evolution can deny that instincts are the vestiges of ancient truths. The question remains whether having spiritual beliefs is instinctive or not. To be considered instinctive, an idea must be common to all humans as a species; otherwise, it could be just an imaginary concept, a product of a delusional mind.

One could easily prove that spiritual beliefs are common among many people in the world, but this is not enough to prove that they are based on ancient truths.

Since the brain has enormous flexibility to learn new ideas, some apparently common ideas could have been learned through indoctrination. To prove that an idea is instinctive, the idea must be common to all humans as a species, independent of one's cultural background and historic context.

Scientists have recently developed techniques to map neuron connections inside the brain (Thomas Mrsic-Flogel, 2011), thus they could potentially identify neuron connections that are common to all human as species. In theory, scientists could prove that an idea is hereditary by showing that it comes from the area of the brain that involves hereditary traits. Unfortunately, this remains a dream for the distant future.

In practice, the only realistic way to know what people think is by having them verbally share their opinions. Unfortunately, language is ambiguous; people rely on a limited number of words and symbols to describe an almost infinite number of concepts.

If the brain did not have a natural capability for generalization, then a functional language would be virtually impossible. Languages would need one word for each and every concept that people could possible think about. In the same way that an artificial neural network is trained to associate different handwritten figures to the same letter, the human mind needs to associate many different ideas to the same word.

## STATISTICS

The ambiguity of human communication can be illustrated with the word *wine*. Depending on the year, type of grape, and geographical location, a certain wine would have its own particular color, flavor, and aroma. People who understand about wine can easily perceive these attributes and predict the possible commercial value of a bottle of wine. Based on human perception, there are huge differences between a good bottle and a bad bottle of wine.

For the purpose of communication, however, the two wines are associated with the same word. People find that similarities between two bottles of wines are more important than the differences they may demonstrate. When scientists try to describe wine using units of measurement, they meet the exact same problem. There are almost infinite mixtures of alcohol, water, and chemical substances that can be associated with the word *wine*.

To describe this type of ambiguity, scientists developed an area of mathematical theory that is known today as statistics. It is believed that statistics were first applied in the early 1300s to census the economy, population, politics, and traditions in medieval cities (Giovanni Villani, ~1300). The mathematical basis, however, was developed in the mid-1600s to calculate the odds of winning a game of chance (Blaise Pascal & Pierre de Fermat, 1654).

## A GAME OF CHANCE

The origin of statistics as a mathematical theory can be traced back to a game of chance. A game of chance, in the context of this story, can be imagined as a group of people taking turns throwing the dice. The results of the dice throws decide how many points each player gets on each turn, and the player that reaches a certain number of points first is the winner of the game and the prize.

In some cases, the game had to be interrupted before any player could win, and when this happened nobody knew how to divide the prize. The simplest solution was to divide the prize proportionally to the players' numbers of points, but this was clearly unfair. If one player was almost sure that he or she was not going to win, then he or she could quit the game and still get a share of the prize.

A much better solution was to divide the prize according to the players' odds of winning at the moment the game was interrupted. To calculate these odds, mathematicians wrote down each and every possible dice roll that each player could have gotten turn after turn. These showed all possible results of the game, and the fraction of results showing a particular winner was his or her probability of winning the game in theory.

## THE PROBABILITY DISTRIBUTION

The simplest way to visualize the probabilities of winning a game of chance is by drawing a probability distribution. In this context, a probability distribution can be imagined as a series of vertical bars in which the bars represent players and the height of the bars represents their probability of winning the game.

When the game starts, all bars are at exactly the same height, thus all players have the same probability to win. As the game proceeds, some players accumulate more points than others, and thus the bar of the winner gradually gets higher, and the bars of the losers gradually get shorter. Finally, when the game ends, there is only one bar with a 100% probability to win.

In a more scientific context, probability distributions are used to visualize any kind of data that can be sorted into groups. In this case, the bars represent the groups and the height of the bars represents the probability of having data within each group.

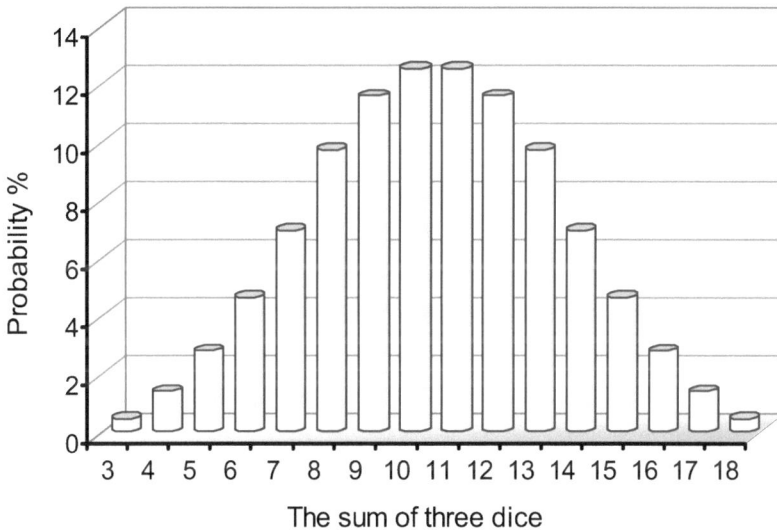

**Figure 2:** My calculation of the probability distribution for the sum of three dice.

One historic example of this concept is the bill of mortality published in the mid-1600s (John Graunt, 1662). The bills of mortality sorted the people who had died according to age groups; thus the height of the bars represented the probability of dying before one's next birthday. Of course, one could say that these people died for many different reasons, but because they lived in the same community, their deaths were connected in one way or another.

## THE REAL VALUE

Another classic example of probability distribution is a student measuring a pendulum with a wrist watch. At first, the student thinks that measuring pendulum swings is a very simple operation. He or she only needs to look at the watch when the pendulum passes through a certain point and then look at the watch again when the pendulum returns exactly to the same place.

Soon, however, the student realizes that it is impossible to look at the pendulum and the watch at the same time. Every time the student measures the pendulum, he or she gets a slightly different result. Theory says that the duration of the swing only depends on the length of the string, but the student does not know which measurement, if any, gave him or her the correct pendulum time.

Mathematicians in the early 1700s realized that the solution to that problem was to measure the pendulum as many times as possible. They figured that random errors were sometimes positive and sometimes negative, thus they tended to cancel each other out. If the same measurement was repeated infinite times, then the average of all measurements was supposed to be equal the real value (Jacob Bernoulli, 1713).

Of course, nobody can repeat the same measurement infinite times, so scientists deduced theoretical probability distribution curves showing how experimental data are normally distributed around the real value (Carl Gauss, 1809). These curves only require a relatively small number of measurements to calibrate the parameters, and then they can be used to withdraw general conclusions as if infinite measurements were taken.

## REPRESENTATIVE SAMPLE

By using probability distribution curves, scientists can draw general conclusions from a relatively small amount of experimental data. The question is how small the number of data can be while still producing a reliable result. If the equipment is defective or the problem was not well understood, then a person could measure the same error a few times and conclude that it is a real value.

For example, if somebody wants to know the chances of a customer to buy a particular bottle of wine, then he or she could go to a liquor store and ask how many bottles of this wine have been sold lately. However, people going to that particular liquor store may have been influenced by a commercial, for example, and the results may not represent what people would usually buy on any given day.

The solution to such inconsistency is the use of representative samples. A representative sample is a number of data large enough to guarantee that all errors are balanced with the real value. In the survey about wine, for example, the effect of commercials could have been canceled out by gathering data in many liquor stores at different times of the year and in different parts of the world.

## A MEASURE OF LOVE

Concepts like love can be statistically measured with representative samples. When one asks people to describe love with words, for example, the replies are usually ambiguous. Some people may say that love is an emotional attachment to another person, while others may say that it is a virtue that represents human kindness. Some people may say that love is a passionate sexual desire, while others may say that love is a spiritual experience. For some, love can simultaneously cause pleasure and pain. For others, love may not even exist.

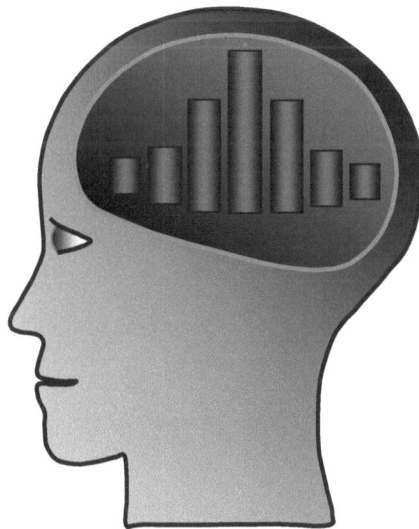

**Figure 3:** My ambiguous representation of a human idea.

For the unbeliever in love, listening to contradictory opinions may confirm that love is a delusion. If something is real, the unbeliever may say, everyone should perceive it in the same way.

For the believer in love, on the other hand, words are just a method of measurement. Believers know that love is real because they can feel it in their minds and bodies, but when they try to measure it, they introduce too much error. As in the example of the pendulum, every time somebody asks for a measure of love, he or she gets a different result.

## CANCELING INDOCTRINATION

Representative samples are widely used in social sciences to study the human mind. As long as scientists can guarantee that they have enough people to balance the cultural background and historic context, then the opinions of a large number of people would tend to be measured close to the most probable meaning.

Even though some people make mistakes or lie when they are asked for their opinions, the mistakes and lies of a large number of people would cancel the opposite, corresponding lies or mistakes. Exactly the same rationale can be applied to spiritual beliefs. Religious indoctrination would certainly skew the opinions toward one belief or another, but it would be counterbalanced by a different indoctrination if many people from totally different cultures are asked about it.

Since indoctrination is supposed to be a man-made illusion, then it would be very unlikely for all societies in different parts of the world and different times in history to have come up with exactly the same illusion. If the source of spiritual beliefs is real, then a probability distribution of all spiritual beliefs would tend to be distributed around the human understanding of that source.

## THE INNER TRUTH

It seems apparent that some spiritual beliefs are common to many different cultures. The belief in life after death, for example, seems to be one of the pillars of many modern religions. With so many religions at different times in history and different places in the world, it is hard to deny that the belief in life after death may be an inner truth.

If life after death were a man-made idea, then believers in it would be much more difficult to find. Just try to imagine what would happen if somebody asked somebody else for money and promised to pay after death. Surprisingly, this is what most organized religions do. Generation after generation of people could not have bought into the promise of life after death without something inside them that wanted to believe it.

Spiritual beliefs can only make sense if they are also based on inner truth. If the theory of evolution is correct, then humans consisted of no more than a single cell at the moment of creation. Humans may have experienced the universe in ways that are unimaginable to the logical person, and part of what they have seen may still be visible inside the human mind.

If one could dig deeper and deeper into these inner truths, one may see the reflection of a reality that is invisible to the senses of perception. Regardless of whether people believe it or not, evolution may not have favored ideas that did not increase the chances of survival and reproduction. If spirituality is an inner truth, then people must have witnessed that truth in a previous life.

Again, it is against the scientific facts to assume that everything that looks real inside the human mind has to look real to the human eyes. Scientific facts and spiritual beliefs can both be real because they were witnessed through different minds. Maybe evolution can explain why so many scientists today do not see any conflicts between science and spirituality.

# CHAPTER 7

## To Believe or Not to Believe

**Abstract:** "To Believe or Not to Believe" presents some final thoughts about a possible link between science and spirituality. After showing that it is possible to be a scientist and still have spiritual beliefs, this chapter asks the question of whether it makes any difference if one has spiritual beliefs or not. To answer that question, the author proposes the analogy of a color-blind person, explaining that sometimes people need to believe in hypothetical concepts in order to understand how these concepts may affect their levels of happiness and chance for success. The chapter concludes with the claim that everybody should seize the opportunity to explore their mystical thoughts.

**Keywords:** Summary of the book, spiritual beliefs, mystical thoughts, invisible elements, imaginary entities, delusion, reasonable doubt, life after death, the end of the world, prayers, supernatural power, higher consciousness, faith, coincidences, skepticism, happiness and success, absolute knowledge, religious texts, religious intolerance, dogmas.

## SUMMING UP

As it was anticipated in the introduction, the objective of this book was to provide a scientific approach to philosophical thinking and encourage readers to explore their spiritual beliefs. This book aimed to inform readers about scientific theories and explain how these theories could be interpreted to validate the mystical and supernatural.

Unlike other similar books that explore the connection between science and religion, this book refrained from manipulating science to fit a particular dogma; it simply highlighted scientific concepts that are remarkably similar to those ideas traditionally associated with spiritual beliefs.

Science shows that there are more elements in the universe than those that are visible to the senses of perception. It demonstrates that, between measurable objects and events, there are invisible areas of the universe through which these concepts connect. It does not matter how far science may advance in the future; it is theoretically impossible to describe this invisible connection by using only scientific units of measurement.

## A REASON TO BELIEVE

This book also took a more speculative look into science. Even though the scientific method can only validate facts that are measurable and predictable, it can also define imaginary entities that are only visible in the mind of the observer. If one looks into those entities with the conviction of a spiritual person, then it is possible to see a connection between science and spirituality.

With some level of speculation, this book showed that the universe may have memory of the past and that life may have consciousness of its own existence. Furthermore, scientific evidence provides reasons to believe that there is a spiritual inner truth inside each and every human being. If one could dig dipper and dipper into this inner truth, one may be able to see the universe from the perspective of a previous life.

However, understanding this inner truth may require accepting that not everything that is known by humans is necessarily reasonable to humans.

In theory, there could be knowledge inside the human mind that does not come from the mind of other humans. The question for the skeptic person is whether this knowledge can answer any question in life that science has not already answered. The answer to that question, however, can only be found in the mind of the observer.

## REASONABLE DOUBT

People may have many reasons for having and not having spirituals beliefs. Some people may find their beliefs unquestionable, while others have reasonable doubt. Those with doubt seek some evidence that is independent of the observer. For them, spiritual beliefs can only be real if they have a real effect on the believers' circumstances. Otherwise, they may say, it is very difficult to differentiate the truth from the fantasies of a delusional mind.

Of course, spiritual beliefs such as life after death or the end of the world are not supposed to be tested during life, but there are other spiritual beliefs that are supposed to have a positive impact on one's circumstances. For example, if somebody prays for peace, then he or she is expecting those prayers to bring some peace to the world.

At a much more practical level, the same is true for a gambler who relies on an amulet to help him or her win a game of chance. The gambler believes that the amulet has the power to change the circumstances of the game. A logical person would assume that, after losing all of his or her money, the gambler would eventually realize that the amulet does not have any power. However, this is not always the case; some gamblers may still rely on the same amulet the next time they play.

This is not necessarily a contradiction. If it is true that prayers and objects have the supernatural power to change circumstances, then they must also have the power to decide when to do it. The power cannot be fully predictable, because otherwise it would have already been validated by science. Prayers and amulets can only increase the chances of something to happen, but it has to be a higher level of consciousness that ultimately makes the choice.

## THE TRUE BELIEVER

The true believer must be able to accept that there is a higher level of consciousness beyond the boundaries of human perception. This higher level of consciousness can behave in mysterious and capricious ways, because it does not necessarily respond to the believer's logic. An excellent example is a religious man who remains faithful in his religion despite many adversities in life.

This religious man may be sincerely convinced that his faith should provide everybody with health and wealth, but at the same time he may be extremely sick and poor. A true believer does not see any contradiction, because he assumes that there is a spiritual reason for being sick and poor at that particular point in life. If circumstances change later and the man becomes healthy and wealthy, then the true believer would probably say that they changed because of his faith.

To the skeptic person, the true believer sounds delusional. It is difficult to understand why anybody would give credit to a supernatural entity for the things in life that he or she had to work hard to obtain. The skeptic would rather give credit to his or her personality and the circumstances for anything good or bad that happens in life. As a rule of thumb, most people would claim that everything good happens because of their efforts while everything bad happens because of the circumstances.

## TO BELIEVE OR NOT TO BELIEVE

Ultimately, anything that people believe or do not believe is reasonable if it increases their levels of happiness and chances of success. Some people may find that they can enjoy life to the fullest without the need for any spiritual beliefs, while others may find that they would never feel complete without a spiritual purpose in life. Without such a purpose, for example, some people may feel that they would never have enough material possessions to stop wanting something else.

Even for the most skeptical scientists in the world, it would be absurd to assume that one has an absolute knowledge of the universe. A real scientist would always assume that there is always something else to discover. One should approach spiritual beliefs with the open mind of a scientist.

For the purposes of analogy with the scientific method, my hypothesis states that human life is connected to a higher level of consciousness, and the circumstances in life are the experimental data that reject or confirm the hypothesis. The only difference with a real scientific hypothesis is that the effect on the circumstances is not necessarily predictable. That is, what some people may perceive as evidence, other people would perceive as coincidence.

## TO NOT SEE TO BELIEVE

A simple analogy that can help one visualize the effect of a consciousness in the circumstances involves trying to imagine the world from the perspective of a colorblind man. There is a rare case of colorblindness in which a person sees everything normally except with no notion of color at all. It is hard to imagine how this man may see the world without color, but one could presume that it would be the equivalent of watching an old black and white movie.

If life were like a black and white movie, then color would have no effect on the way this man sees the world. On the other hand, color would have an effect on the way the world sees the man. This is because the colorblind man shares the world with other people who are able to see color.

For example, if the man were to wear a dark suit to a formal meeting, he may elicit respect and consideration from others, but if he wore a red suit in the same situation, others may treat him with irony or contempt. Although color is imperceptible to his senses, he may still perceive them in certain circumstances because they exist in the minds of his observers.

The only way for this man to understand the effect of color in life is to look at the behavior of other people. The first step, of course, is to believe that colors are real. Once the colorblind man accepts that colors may have an effect in the minds of other people, then he can try to create a mental picture of the colors as reflected upon the circumstances. Eventually, the colorblind man would be able to see colors by using his imagination rather than his senses of perception.

## THE SPIRITUAL EXPERIENCE

To see colors, a colorblind man needs to create an imaginary connection between colors and people. However, creating this connection is not a straightforward scientific problem, because people's behaviors are not necessarily consistent in different contexts. For example, wearing a red suit to an informal party could elicit attention, while wearing the same suit in a formal party can make him to lose his job.

People have the power to make decisions; thus with exactly the same action one could elicit a totally different reaction. The connection between action and reaction in social life is not a mathematical law, but a distribution of human experience showing the probability for something to happen.

Exactly the same rationale applies to spiritual beliefs. If there really is a connection between humans and a higher level of consciousness, then people should perceive that connection as the probability for something to happen. Spiritual beliefs should not give people the power to go against the laws of nature, but rather to take advantage of a universe that other people see as full of coincidences.

## RELIGIOUS TEXTS

The obvious difference between a colorblind man and a spiritual person is that the colorblind man can ask other people to explain colors to him. The source of spiritual beliefs, on the other hand, is invisible to all humans. People cannot simply ask other people to describe the source of spiritual beliefs.

For this reason, religious books represent an invaluable source of spiritual wisdom. It is well known that most civilizations were organized around spiritual beliefs, and their religious texts are probably a reflection of those religious experiences. Metaphorically speaking, reading a religious book is like looking at the source of spiritual beliefs through the collective mind of many other people who have seen the same thing.

It is true that many religious people see these books as collections of revelations, but I also believe that even ideas must also go through some form of evolution. If one accepts that there are different ways to connect with the source of spiritual beliefs, then it would be reasonable to assume that some ways are more effective than others. False spiritual beliefs cannot pass effectively from person to person and generation to generation.

## RELIGIOUS INTOLERANCE

It is also naïve to reject religious books based on the literal meaning of their words. The ultimate objective of religious texts is to connect people with the subject of spiritual beliefs. If the source is real, then it does not matter whether or not the words are reasonable in their literal interpretation. The objective is to inspire people to understand the universe from the perspective of a higher consciousness.

Unfortunately, some religious books were written for specific communities of people living in other parts of the world and other periods in history. Like in any other community around the world, people expected other people to speak, eat, dress, and behave similarly. By sharing the same customs and rules, people create the illusion that they are connected through the same common idea.

For this reason, one should be open minded enough to differentiate spirituality from dogma. Different religious communities may have developed contradictory dogmas and moral rules, but this does not mean that the source of spiritual beliefs is also contradictory. One hopes that, as humanity evolves, all people would eventually connect through the same common idea.

## CONCLUSIONS

If the source of spiritual beliefs is real, then it must be a higher level of consciousness. Only an invisible consciousness can be invisible to the scientific method, and at the same time evident to human beings. If people want to see this consciousness through its effect on the circumstances, then they first have to believe it.

Many people in the world claim to have spiritual beliefs that would be statistically absurd to assume to be scientifically impossible. Even the most skeptic person in the world should explore his or her mystical thoughts with the open mind of a scientist. Life is so full of good and bad experiences that sooner or later one should have enough experimental data to accept or reject this hypothesis.

The goal, of course, should be to increase one's happiness and chances for success. Despite the apparently irrational dogmas imposed by some religions books, there is no doubt that religious people feel connected to a higher level of consciousness. In one way or another, most spiritual people would agree that their faith has the power to affect their circumstances.

Of course, many people in the world feel happy and successful without any spiritual connection. It is obvious that these people have already developed a positive connection with life, and they may not need any supernatural entity to change it. For these people, life is good and they will live it to its fullest.

Many other people, however, go through periods of extreme pain and despair. It is under these physiological threats to survival that the mind intuitively looks to that inner truth. Although unfortunate, extreme pain and despair are sometimes the best driving forces for finding purpose in life.

Of course, suffering is cruel. Extreme pain and despair may induce people to look for miraculous solutions, and in many cases, try change reality by using drugs and alcohol. The solution, therefore, is rather simple: we should search for inner truth when life is good and the mind is clear.

We must develop a positive connection with the universe before we need to change our circumstances in life. When the bad times come, one should know with a surety that there is a good reason for being alive. If one, on the other hand, can never find purpose in life, then the conclusion is still good − we must accept life as it is.

# REFERENCES

Adam Riess, 1998: Riess A.G. *et al.* (1998). Observational evidence from supernovae for an accelerating universe and a cosmological constant. *The Astronomical Journal*, 116, 1009-1038.

Adolf Hitler, 1925: Hitler, A. (1938). *Mein kampf.* London: Hurst & Blackett.

Albert Einstein, Jun 1905: Einstein, A. (1905). On a heuristic viewpoint concerning the production and transformation of light (*Annalen der Physik*, 17(6)., 132-148). Retrieved January 1, 2012, from http://users.physik.fu-berlin.de/~kleinert/files/eins_lq.pdf

Albert Einstein, Sep 1905: Einstein, A. (1905). On the electrodynamics of moving bodies (*Annalen der Physik*, 17(10)., 891-921). Retrieved January 1, 2012, from http://users.physik.fu-berlin.de/~kleinert/files/eins_specrel.pdf

Albert Einstein, Nov 1905: Einstein, A. (1905). Does the inertia of a body depend upon its energy content? (*Annalen der Physik*, 18(13)., 639-641). Retrieved January 1, 2012, from http://users.physik.fu-berlin.de/~kleinert/files/e_mc2.pdf

Albert Einstein, 1916: Einstein, A. (1997). The foundation of the general theory of relativity. In *The collected papers of Albert Einstein* (Vol. 6, pp. 146-200). Princeton, NJ: Princeton University Press.

Albert Michelson & Edward Morley, 1887: Michelson, A.A. & Morley, E.W. (1887). On the relative motion of the earth and the luminiferous ether. *American Journal of Science*, 34, 333-345.

Alessandro Volta, 1800: Giuliano, P. (2003). *Volta, science and culture in the age of enlightenment.* Princeton, NJ: Princeton University Press.

Alister McGrath, 2004: McGrath, A.E. (2004). *The twilight of atheism: The rise and fall of disbelief in the modern world.* New York: Doubleday.

André-Marie Ampère, 1820: Ampère, A.-M. (1826). *Théorie des phénomènes électro-dynamiques, uniquement déduite de l'expérience.* Paris: Méquignon-Marvis.

Antoine Lavoisier, ~1750: Moore, F.J. (1918). *A history of chemistry.* New York: McGraw-Hill.

Antoine Lavoisier, 1783: Lavoisier, A.L. (1783). *Réflexions sur le phlogistique, pour servir de développement à la théorie de la combustion & de la calcinations.* Paris: Académie des sciences.

Archimedes of Syracuse, ~250 BC: Rowland, D. & Howe, T.N. (Eds.). (1999). *Vitruvius. Ten books on architecture.* Cambridge, UK: Cambridge University Press.

Aristotle's Generation, ~350 BC: Peck, A.L. (Ed.). (1979). *Generation of animals.* Cambridge, Massachusetts: Harvard University Press.

Aristotle's Heavens, ~350 BC: Stocks J.L. & Joachim, H.H. (Eds.). (2006). *On the heavens and on generation and corruption.* Digireads.com Publishing. Retrieved January 1, 2012, from http://www.digireads.com

Aristotle's Physics, ~350 BC: Hope, R. (Ed.). (1961). *Aristotle's physics: With an analytical index of technical terms.* Lincoln: University of Nebraska Press.

Benjamin Franklin, 1752: Franklin, B. (1752). A letter of Benjamin Franklin, Esq; to Mr. Peter Collinson, F.R.S. concerning an electrical kite. *Philosophical Transactions*, 47, 565-567. Retrieved January 1, 2012, from http://rstl.royalsocietypublishing.org/content/47/565.full.pdf

Benjamin Thompson, 1798: Thompson, B. (1798). An inquiry concerning the source of heat which is excited by friction. *Philosophical Transactions*, 18, p. 286.

Bhaskara Achārya, ~1150: White, L.T. (1978). *Medieval religion and technology: Collected essays*. Berkeley, California: University of California Press.

Blaise Pascal & Pierre de Fermat, 1654: David, H.A. & Edwards, A.W.F. (2001). *Annotated readings in the history of statistics*. New York: Springer.

Camillo Golgi, 1873: Mazzarello, P. (2009). *Golgi: A biography of the founder of modern neuroscience*. Oxford: Oxford University Press.

Carl Gauss, 1809: Gauss, C.F. (1809). *Theoria motus corporum coelestium in sectionibus conicis solem ambientium*. Ghent, Belgium: Ghent University.

Carl Sagan, 1973: Sagan, C. (1973). *Communication with extraterrestrial intelligence (CETI)*. Cambridge, Massachusetts: MIT Press.

Charles Darwin, 1859: Darwin, C. (1859). *On the origin of species by means of natural selection, or the preservation of favoured races in the struggle for life*. London, UK: John Murray.

Christiaan Huygens, 1656: Huygens, C. (1658). The timepiece. *Antiquarian Horology*, 7(1). Retrieved January 1, 2012, from http://www.nawcc-index.net/Articles/Huygens_Horologium_English.pdf

Christiaan Huygens, 1678: Huygens, C. (1690). *Traité de la lumiere*. Leiden, Netherlands: Pieter van der Aa.

Claude Shannon, 1948: Shannon, C.E. (1948). A mathematical theory of communication. *Bell System Technical Journal*, 27, 379-423 & 623-656.

Claudius Galenus, ~150: Linacre, T. (Ed.). (1881). *Galeni pergamensis de temperamentis: et De inaequali intemperie libri tres*. Cambridge, UK: Cambridge.

Claudius Ptolemy, ~150 AD: Toomer, G.J. (Ed.). (1998). *Ptolemy's Almagest*. Princeton, NJ: Princeton University Press.

Copenhagen interpretation, 1955: Heisenberg, W. (1958). *Physics and philosophy: The revolution in modern science*. New York: Harper.

Craig Venter, 2007: Gibson, D.G. *et al.* (2010). Creation of a bacterial cell controlled by a chemically synthesized genome. *Science*, 329(5987)., 52-56.

Daniel Bernoulli, 1738: Bernoulli, D. & Bernoulli, J. (1968-reprint). *Hydrodynamics and Hydraulics*. New York: Dover Publications.

David Cohen, 1968: Cohen, D. (1968). Magnetoencephalography: Evidence of magnetic fields produced by alpha rhythm currents. *Science*, 161, 784-786.

Denis Papin, 1679: Lavoisier, A. (1984-reprint). *Elements of chemistry*. New York: Courier Dover Publications, Inc.

Diego de Landa, 1566: Gates, M. (Ed.). (1978). *Yucatan before and after the conquest*. New York: Dover Publications, Inc.

Donald Hebb, 1949: Hebb, D.O. (1949). *The organization of behavior: A neuropsychological theory*. New York: John Wiley & Sons, Inc.

Douglas Theobald, 2010: Theobald, D.L. (2010). A formal test of the theory of universal common ancestry. *Nature*, 465, 219-222.

Ecclesiastes 1:5: "The sun rises and the sun sets, and hurries back to where it rises".

Edwin Hubble, 1929: Hubble, E. (1929). A relation between distance and radial velocity among extra-galactic nebulae. *Proceedings of the National Academy of Science*, 15, 168-173.

Elaine Ecklund, 2010: Ecklund, E.H. (2010). *Science vs. religion: What scientists really think*. Oxford, UK: Oxford University Press.

Epperson *vs.* Arkansas, 1968: Susan Epperson *et al.* v. Arkansas, 393 U.S. 97 (1968).

Ernest Nichols & Gordon Hull, 1903: Nichols, E.F. & Hull, G.F. (1903). The pressure due to radiation. *The Astrophysical Journal*, 17(5)., 315-351.

Ernest Rutherford, 1911: Rutherford, E. (1911). The scattering of α and β particles by matter and the structure of the atom. *Philosophical Magazine*, 6(21)., 669-688.

Erwin Schrödinger, 1926: Schrödinger, E. (1926). An undulatory theory of the mechanics of atoms and molecules. *Physical Review*, 28(6)., 1049-1070.

Erwin Schrödinger, 1935: Trimmer, J.D. (Ed.). (1983). The present situation in quantum mechanics: A translation of Schrödinger's "cat paradox paper". *Proceedings of the American Philosophical Society*, 124, 323-338.

Evangelista Torricelli, 1643: Toscano, F. (2008). *L'erede di Galileo: Vita breve e mirabile di Evangelista Torricelli*. Milano, Italy: Sironi Editore.

Felisa Wolfe-Simon, 2011: Wolfe-Simon, F. *et al.* (2011). A bacterium that can grow by using arsenic instead of phosphorus. *Science*, 332, 1163-1166.

First Amendment, 1791: National Archives and Records Administration. *The charters of freedom*. Retrieved        January        1,        2012,        from http://www.archives.gov/exhibits/charters/bill_of_rights_transcript.html

Francesco Redi, 1668: Bigelow, M. & Bigelow, R.P. (Eds.). (1909). *Experiments on the generation of insects*. Chicago, US: The Open Court Publishing Company.

Friedrich Miescher, 1869: Dahm, R. (2004). Friedrich Miescher and the discovery of DNA. *Developmental Biology*, 278, 274-288.

Friedrich Wöhler, 1828: Kinne-Saffran, E. & Kinne, R.K.H. (1999). Vitalism and synthesis of urea: From Friedrich Wöhler to Hans A. *Krebs. Am J Nephrol*, 19, 290-294.

Fritz Zwicky, 1933: Zwicky, F. (1937). On the masses of nebulae and of clusters of nebulae. *Astrophysical Journal*, 86, p. 217.

Galileo Galilei, 1590: Settle, T.B. (1983). Galileo and early experimentation. In *Springs of scientific creativity: Essays on founders of modern science* (pp. 3-20). Minneapolis: University of Minnesota Press.

Galileo Galilei, 1602: Drake, S. (1978). *Galileo at work: His scientific biography*. Chicago: University of Chicago Press.

Galileo Galilei, 1610: Van Helden, A. (Ed.). (1989). *The Sidereal Messenger*. Chicago: The University of Chicago Press.

Galileo Galilei, 1638: Henry C. & de Salvio, A. (Eds.). (1954). *Dialogues Concerning Two New Sciences*. New York: Dover Publications, Inc.

General Conference on Weights and Measures, 1967: Resolutions of the 13th meeting of the CGPM.        Retrieved        January        1,        2012,        from http://www.bipm.org/jsp/en/ListCGPMResolution.jsp?CGPM=13

Genesis 1:27: "So God created mankind in his own image, in the image of God he created them; male and female he created them".

Giordano Bruno, 1584: Singer, D.W. (1950). *Giordano Bruno, his life and thought: With annotated translation of his work; On the infinite universe and worlds*. New York: Henry Schuman, Inc.

Giovanni Villani, ~1300: Villani, G. (2010-reprint). *Nuova cronica*. Bologna, Italy: Zanichelli editore S.p.A.

Gregor Mendel, 1865: Mendel, G. (1866). Experiments in plant hybridization (*Ver. Brünn*, 4, 3-47). Retrieved January 1, 2012, from http://www.mendelweb.org/Mendel.html

Gustave Coriolis, 1829: Coriolis, G. (1829). *Du calcul de l'effet des machines, ou Considérations sur l'emploi des moteurs et sur leur évaluation, pour servir d'introduction a l'étude spéciale des machines*. Paris, France: Carilian-Goeury.

Hans Berger, 1924: Haas, L.F. (2003). Hans Berger (1873-1941)., Richard Caton (1842-1926). and electroencephalography. *Journal of Neurology, Neurosurgery & Psychiatry,* 74(1)., p. 9.

Hans Driesch, 1914: Driesch, H. (1914). *The history and theory of vitalism.* London, UK: Macmillan and Co., Ltd.

Hans Ørsted, 1820: Jelved, K., Jackson, A.D. & Knudsen, O. (Eds.). (1997). *Selected scientific works of Hans Christian Ørsted.* Princeton, NJ: Princeton University Press.

Heinrich Hertz, 1886: Thomson, J.J. (2010). *Notes on recent researches in electricity and magnetism: Intended as a sequel to professor Clerk-Maxwell's treatise on electricity and magnetism.* Cambridge, UK: Cambridge University Press.

Hendrik Lorentz, 1899: Lorentz, H.A. (1904). Electromagnetic phenomena in a system moving with any velocity smaller than that of light. *Proceedings of the Royal Netherlands Academy of Arts and Sciences,* 6, 809-831.

Henrietta Leavitt, 1912: Henrietta, L.S. & Edward, P.C. (1912). Periods of 25 variable stars in the small magellanic cloud. *Harvard College Observatory Circular,* 173, p. 1.

Hippolyte Fizeau, 1849: Poincaré, H. (1904). Experiments of MM. Fizeau and Gounelle. In *Maxwell's theory and wireless telegraphy: Part 1. Maxwell's theory and Hertzian oscillations* (pp. 52). New York: McGraw Publishing Company.

Hugh Everett, 1957: Everett, H. (1957). Relative state formulation of quantum mechanics. *Reviews of Modern Physics,* 29, 454-462.

Hugo de Vries & Carl Correns, 1900: Bowler, P.J. (2003). *Evolution: the history of an idea.* Berkeley, US: University of California Press.

Isaac Newton, ~1660: Stukeley, W. (1752). *Memoirs of Sir Isaac Newton's life.* The Royal Society, Retrieved January 1, 2012, from http://ttp.royalsociety.org/silverlight/?id=1807da00-909a-4abf-b9c1-0279a08e4bf2

Isaac Newton, 1675: Newton, I. (1979-reprint). *Opticks: Or a treatise of the reflections, refractions, inflections & colours of light.* New York: Courier Dover Publications.

Isaac Newton, 1687: Motte, A. (Ed.). (1729). *The mathematical principles of natural philosophy.* London, UK: Printed for B. Motte.

Isidor Rabi, 1930: Lombardi, M.A., Heavner, T.P. & Jefferts, S.R. (2007). NIST primary frequency standards and the realization of the SI second. *Journal of Measurement Science,* 2(4)., p. 74.

Jacob Bernoulli, 1713: Sheynin, O. (Ed.). (2005). On the law of large numbers (*The art of conjecturing; part four showing the use and application of the previous doctrine to civil, moral and economic affairs*). Retrieved Janaury 1, 2012, from http://www.sheynin.de/download/bernoulli.pdf

James Joule, 1845: Joule, J.P. (1850). *On the mechanical equivalent of heat. Philosophical Transactions of the Royal Society of London,* 140(1)., 61-82.

James Maxwell, 1859: Mahon, B. (2003). *The man who changed everything: The life of James Clerk Maxwell.* Chichester, UK: John Wiley & Son, Ltd.

James Maxwell, 1862: Maxwell, J.C. (1865). A dynamical theory of the electromagnetic field. *Philosophical Transactions of the Royal Society of London,* 155, 459-512.

James Maxwell, 1874: Maxwell, J.C. (1873). *A treatise on electricity and magnetism.* London, UK: Macmillan & Co.

James Watson & Francis Crick, 1953: Watson, J.D. & Crick, F.H.C. (1953). A structure for deoxyribose nucleic acid. *Nature*, 171(4356)., 737-738.

James Watt, 1765: Watt, J. (1769). A method of lessening the consumption of steam in steam engines-the separate condenser. UK patent no. 913.

Jean Cuvier, 1830: Coleman, W. (1996). *Georges Cuvier, zoologist: A study in the history of evolution theory*. Cambridge, Massachusetts: Harvard University Press.

Jean Lamarck, 1809: De Monet de Lamarck, J.B.P.A. (2011-reprint). *Philosophie zoologique: Ou exposition; des considerations relative à l'histoire naturelle des animaux*. Cambridge, UK: Cambridge University Press.

Johann Bessler, 1717: Gould, R.T. (2003-reprint). Orffyreus's wheel. In *Oddities: A book of unexplained facts* (pp. 89-116). Whitefish, Montana, US: Kessinger Publishing.

John Graunt, 1662: Willcox, W.F. (1939-reprint). *Natural and political observations made upon the bills of mortality*. Baltimore, Maryland: The Johns Hopkins press.

John Whitcomb & Henry Morris, 1961: Whitcomb, J.C. & Morris, H.M. (1961). *The genesis flood: The biblical record and its scientific implications*. Philadelphia, US: Presbyterian & Reformed Publishing.

Joseph Gay-Lussac, 1802: Gay-Lussac, J.-L. (1935). The expansion of gases by heat. In *A source book in physics*. New York: McGraw-Hill.

Joseph Hafele & Richard Keating, 1971: Hafele, J. & Keating, R. (1972). Around the world atomic clocks predicted relativistic time gains. *Science*, 177, 166-168 & 168-170.

Judge Jones III, 2005: Case 4:04-cv-02688-JEJ, Document 342. Retrieved January 1, 2012, from http://www.pamd.uscourts.gov/kitzmiller/kitzmiller_342.pdf

Jules Poincaré, 1900: Poincaré, J.H. (1913). *The Foundations of science: Science and hypothesis, the value of science, science and method*. New York: The Science Press.

Kitzmiller *et al. vs.* Dover, 2005: Tammy Kitzmiller *et al.* v. Dover Area School District *et al.*, 400 F. Supp. 2d 707 (M.D. Pa. 2005).

Kurt Gödel, 1949: Gödel, K. (1950). Rotating universes in general relativity theory. *Proceedings of the international congress of mathematicians in Cambridge*, 1, 175-181.

Lord Rayleigh & Sir James Jeans, 1905: Jeans, J.H. (1905). On the partition of energy between matter and æther. *Philosophical Magazine*, 10(55)., p. 91.

Louis de Broglie, 1924: de Broglie, L. (1924). *Recherches sur la théorie des quanta (Research on the quantum theory)*. Thesis: University of Paris, France.

Louis Pasteur, 1859: Robbins, L. (2001). *Louis Pasteur and the hidden world of microbes*. New York: Oxford University Press.

Ludwig Boltzmann, 1877: Gallavotti, G., Reiter, W.L. & Yngvason, J. (2008). *Boltzmann's legacy*. Zürich, Zwitzerland: European Mathematical Society.

Luigi Galvani, 1791: Green, R.M. (Ed.). (1953). *Commentary on the effect of electricity on muscular motion*. Cambridge, MS: Elizabeth Licht Publisher.

Max Planck, 1900: Planck, M. (1967). On the theory of the energy distribution law of the normal spectrum. In *The old quantum theory* (pp. 82). Oxford, UK: Pergamon Press.

Maynard Smith, 1978: Smith, M.J. (1978). *The evolution of sex*. Cambridge, MS: Cambridge University Press.

Michael Faraday, 1831: Faraday, M. (1855). *Experimental researches in electricity*. London, UK: Richard Taylor and William Francis.

Monsignor Lemaître, 1931: Lemaître, G. (1931). Expansion of the universe, the expanding universe. *Monthly Notices of the Royal Astronomical Society*, 91, 490-501.

NASA: NASA Science/Astrophysics: Dark energy, dark matter. Retrieved January 1, 2012, from http://science.nasa.gov/astrophysics/focus-areas/what-is-dark-energy

Nicolas Carnot, 1824: Carnot, N. (2009-reprint). *Reflections on the motive power of fire: And other papers on the second law of thermodynamics by É. Clapeyron and R. Clausius.* Mineola, NY: Dover Publisher.

Nicolaus Copernicus, 1543: Charles, G.W. (Ed.). (1995). *On the revolutions of the heavenly spheres.* New York: Prometheus Books.

Niels Bohr, 1913: Bohr, N. (1913). On the constitution of atoms and molecules, part I. *Philosophical Magazine*, 26, 1-24.

Otto Hahn, 1938: Frisch, O.R. (1939). Physical evidence for the division of heavy nuclei under neutron bombardment. *Nature*, 143 (3616)., p. 276.

Otto Loewi, 1921: Loewi, O. (1921). Über humorale übertragbarkeit der herznervenwirkung. *Pflügers Archiv*, 189, 239-242.

Otto von Guericke, 1663: Schiffer, M.B. (2003). *Bringing the lightning down: Benjamin Franklin and electrical technology in the age of enlightenment.* Berkeley, US: University of California Press.

Papal Condemnation of Galileo, 1633: de Santillana, G. (1955). *The crime of galileo.* Chicago, US: University of Chicago Press.

Percival Davis & Dean Kenyon, 1989: Davis, P. & Kenyon, D.H. (1989). *Of pandas and people: The central question of biological origins.* Dallas, US: Haughton Pub. Co.

Piltdown Man, 1912: Walsh, J.E. (1996). *Unraveling Piltdown: The science fraud of the century and its solution.* New York: Random House.

Plato, ~350 BC: Cooper, J.M. and Hutchinson, D.S. (Eds.). (1997). *Plato: Complete works.* Indianapolis, Indiana: Hackett Publishing, Inc.

Psalm 104:5: "He set the earth on its foundations, so that it should never be moved".

René Descartes, 1637: Lafleur L.J. (Ed.). (1960). *Discourse on method and meditations.* New York: The Liberal Arts Press.

Richard Dawkins, 2006: Dawkins, R. (2006). *The God delusion.* New York: Houghton Mifflin Harcourt.

Robert Hooke, 1665: Gunther, R.T. (Ed.). (1961). *Micrographia; or, some physiological descriptions of minute bodies made by magnifying glasses: With observations and inquiries thereupon.* Mineola, US: Dover Publications.

Rudolf Clausius, 1865: Clausius, R. (1865). *The mechanical theory of heat - with its applications to the steam engine and to physical properties of bodies.* London, UK: John van Voorst.

Rudolf Virchow & Robert Remak, 1855: Magner, L.N. (2002). *A history of the life sciences.* New York: Marcel Dekker, Inc.

Santiago Ramón y Cajal, 1891: DeFelipe, J. & Jones, E.G. (Eds.). (1988). *Cajal on the cerebral cortex: An annotated translation of the complete writings.* Oxford, UK: Oxford University Press.

SI Base Units: Bureau International des Poids et Mesures. Retrieved January 1, 2012, from http://www.bipm.org/en/si/base_units

Simon Newcomb, 1895: Newcomb, S. (1895). Tables of the motion of the earth on its axis and around the sun. U.S. Nautical Almanac Office. *Astronomical paper*, 6(1)., 1-169.

Sir Arthur Eddington, 1919: Dyson, F.W., Eddington, A.S. & Davidson, C.R. (1920). A determination of the deflection of light by the sun's gravitational field, from observations made at the solar eclipse of May 29, 1919. *Phil. Trans. Roy. Soc. A.*, 220(571-581)., 291-333.

Socrates, ~450 BC: Cooper, J.M. & Hutchinson, D.S. (Eds.). (1997). *Plato: Complete works.* Indianapolis, Indiana: Hackett Publishing, Inc.

Stanley Miller & Harold Urey, 1953: Miller, S.L. & Urey, H.C. (1959). Organic compound synthesis on the primitive earth. *Science*, 130(3370)., 245.

Stanley Prusiner, 1982: Prusiner, S.B. (1982). Novel proteinaceous infectious particles cause scrapie. *Science*, 216(4542)., 136-144.

Stuart Kauffman, 1993: Kauffman, S. (1993). *The origins of order: Self-organization and selection in evolution.* Oxford, UK: Oxford University Press.

Thales of Miletus, ~600 BC: Hett, W.S. (Ed.). (1957). *On the soul.* Cambridge, Massachusetts: Harvard University Press "Loeb Classical Library".

Thane Heins, 2002: Heins, T.C. (2005). Infinity generator. Canada patent application no. CA20032437745.

The Manhattan Project, 1941: Vincent, J. (1985). *Manhattan: The army and the atomic bomb.* Washington, DC: United States Army Center of Military History.

The National Center for Science Education, 2007: The National Center for Science Education. (2007). *Ten significant court decisions regarding evolution/creationism.* Retrieved January 1, 2012, from http://ncse.com/taking-action/ten-major-court-cases-evolution-creationism

The State of Tennessee *vs.* Scopes, 1925: State of Tennessee v. John Thomas Scopes, 152 Tenn. 424 (Tenn. 1925).

Theodor Schwann & Matthias Schleiden, 1837: Tavassoli, M. (1980). The cell theory: A foundation to the edifice of biology. *Am. J. Pathol.*, 98(1)., p. 44.

Thomas Edison, 1879: Edison, T. (1880). Electric-lamp. US patent no. 223898.

Thomas Jefferson, 1802: The Library of Congress. (1998). *Jefferson's letter to the Danbury Baptists.* Retrieved January 1, 2012, from http://www.loc.gov/loc/lcib/9806/danpre.html

Thomas Morgan, 1910: Morgan, T.H., Sturtevant, A.H., Muller H.J. & Bridges C.B. (1915). *The mechanism of mendelian heredity.* New York: Henry Holt.

Thomas Mrsic-Flogel, 2011: Ko, H. *et al.* (2011). Functional specificity of local synaptic connections in neocortical networks. *Nature*, 473, 87-91.

Thomas Newcomen, 1712: Rolt, L.T.C. (1963). *Thomas Newcomen. The prehistory of the steam engine.* Dawlish, UK: David & Charles.

Thomas Young, 1803: Young, T. (1804). Experiments and calculations relative to physical optics. *Philosophical Transactions of the Royal Society of London*, 94, 1-16.

Thomas Young & Hermann von Helmholtz, 1850: Southall, J.P.C. (Ed.). (1962). *Treatise on physiological optics.* New York: Dover Publications.

USPTO: The United State Patent and Trademark Office: What can and cannot be patented?. Retrieved January 1, 2012, from http://www.uspto.gov/inventors/patents.jsp

Vesto Slipher, 1912: Slipher, V. (1912). The radial velocity of the Andromeda nebula. *Lowell Observatory Bulletin*, 1, 56-57.

Walter Pitts & Warren McCulloch, 1943: McCulloch, W. & Pitts, W. (1943). A logical calculus of ideas immanent in nervous activity. *Bulletin of Mathematical Biophysics*, 5, 115-133.

Walter Sutton & Theodor Boveri, 1902: Baltzer, F. (1964). Theodor Boveri. *Science*, 144, 809-815. / McKusick, V.A. (1960). Walter S. Sutton and the physical basis of Mendelism. *Bull. Hist. Med.*, 34, 487-497.

Walther Flemming, 1882: Flemming, W. (1882). *Zellsubstanz, kern und zelltheilung*. Leipzig, Germany: Verlag Von F.C.W. Vogel.

Werner Heisenberg, 1927: Heisenberg, W. (1930). *The physical principles of quantum theory*. New York: Dover Publications.

William Gilbert, 1600: Thompson, S.P. (Ed.). (1958-reprint). *On the magnet*. New York: Basic Books, Inc.

Wujing Zongyao, 1044: Needham, J. & Ronan, C.A. (1986). Magnetism and electricity. In *The shorter science and civilization in China: An abridgement of Joseph Needham's original text*. (Vol. 3, Chapter 1). Cambridge, UK: Cambridge University Press.

# SUBJECT INDEX

**V**
Vacuum · 33, 36
horror · 33, 35, 36
Vapor · 51, 69
Viruses · 62
Vital
energy · 58, 70
functions · 80
signs · 57
Vitalism · 58, 70
**W**
Water pump · 36, 51
Wavelength · 25
Wave-particle · 28, 29, 58, 74
Waves · 20, 21, 22, 25, 27, 28, 29, 42
function · 28, 30
White light · 26
Wine · 82, 85
**X**
X-ray · 65
**Y**
Year · 48, 49, 50, 85

# AUTHOR INDEX

www.ingramcontent.com/pod-product-compliance
Lightning Source LLC
Chambersburg PA
CBHW041714210326
41598CB00007B/647